WOOD
MAGAZINE®

BUILD YOUR OWN
SHOP JIGS
& FIXTURES

Sterling Publishing Co., Inc.
New York

Library of Congress Cataloging-in-Publication Data

Wood magazine build your own shop jigs & fixtures / from editors of *WOOD*® magazine.

p. cm.

Includes index.

ISBN-13: 978-1-4027-2043-7

ISBN-10: 1-4027-2043-2

1. Woodworking tools--Design and construction.

2. Woodwork--Equipment and supplies--Design and construction.

3. Jigs and fixtures--Design and construction. I. Wood magazine.

TT186.W62 2006

684'.08--dc22

2006004884

Edited by Peter J. Stephano

Designed by Christine Swirnoff

10 9 8 7 6 5 4 3 2 1

Published by Sterling Publishing Co., Inc.

387 Park Avenue South, New York, NY 10016

This edition is based on material found in *WOOD*® magazine articles

© 2007 by *WOOD*® magazine editors

Distributed in Canada by Sterling Publishing

c/o Canadian Manda Group, 165 Dufferin Street

Toronto, Ontario, Canada M6K 3H6

Distributed in the United Kingdom by GMC Distribution Services

Castle Place, 166 High Street, Lewes, East Sussex, England BN7 1XU

Distributed in Australia by Capricorn Link (Australia) Pty. Ltd.

P.O. Box 704, Windsor, NSW 2756, Australia

Printed in China

All rights reserved

Sterling ISBN-13: 978-1-4027-2043-7

ISBN-10: 1-4027-2043-2

For information about custom editions, special sales, premium

and corporate purchases, please contact

Sterling Special Sales Department at 800-805-5489 or

specialsales@sterlingpub.com.

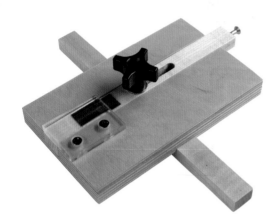

BUILD YOUR OWN
SHOP JIGS
& FIXTURES

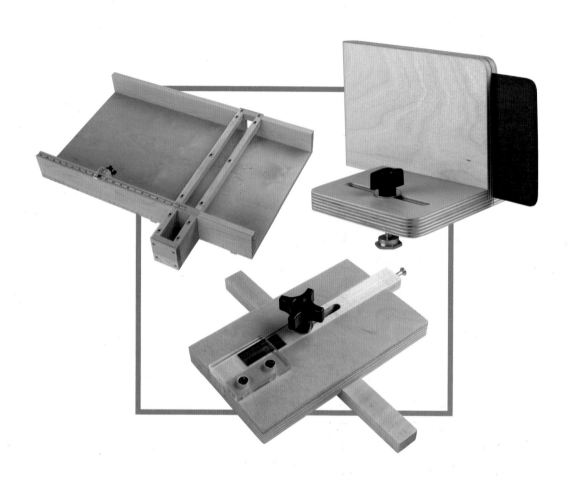

CONTENTS

❖ CHAPTER 4

Bandsaw Aids 98

❖ CHAPTER 5

Drill-Press Options   123

❖ CHAPTER 6

Sanding Solutions 148

❖ CHAPTER 7

Clamping Aids   155

❖ CHAPTER 8

Assorted Shop Aids   166

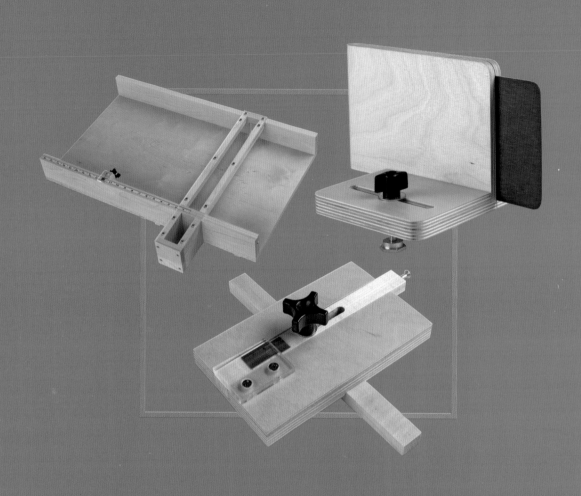

CHAPTER 1

Shop-Made Jigs, Fixtures, and Accessories

IN WOODWORKING JARGON, A "JIG" *is a device that holds a workpiece or tool so that a specific task can be performed more accurately and efficiently. Decades ago, jigs were considered to be somewhat temporary, much like a shim used to momentarily raise a door or cabinet before fastening it in place. "Fixtures" were defined as more refined woodworking aids to keep on hand when their need resurfaced. "Accessories" implied sturdier, often elaborate, add-ons that increased a tool's capacity and efficiency.*

Today's woodworker, though, views them all as like items. No matter their original definitions, jigs, fixtures, and accessories make woodworking easier, faster, and more fun to do because with them the results are predictable and repeatable. In this book, you'll find dozens upon dozens of shop-proven ones that—no matter their size or complexity—were designed with that goal (as well as safety) in mind. Take the time right now to look through this chapter and introduce yourself to what lies ahead. And remember, each project has all the step-by-step instructions, plans, parts views, and list of materials you need to build them.

On *page 38*, we'll show and tell you how to build the dead-on jig shown in **1–2** that enables you to rout through and stopped dadoes, as well as grooves. It's easy to align and set to cut the exact shelf thickness you need. That means no more ruined projects because your cuts were just a bit off.

Want to get your router going in circles? Try the trammel you'll find on *page 52*, and which is shown in **1–3**. You can make it for just a few dollars and an hour's worth of your time, which

This multi-featured router table makes accurate setups quicker and easier. Its portability allows you to use it anywhere in your shop. For full instructions and plans to build it, turn to page 22.

The secret to this dado jig is a flush-trimming pattern bit with a bearing above the cutters. See it on page 38.

1–2.

In Chapter 2, "Router Accessories," you'll see a baker's dozen worth of devices that you can make to get maximum capability from this already quite versatile power tool. For instance, the benchtop router table, pictured in **1–1**, features a quick-locking fence, levelers that ensure a perfectly aligned tabletop, a dust-collection port, and lightweight yet sturdy construction. And at a cost of $100 or less (plus the wood), it won't break your budget!

1–3.

Turn your router into a compass with this jig. See page 52.

means you'll be quickly cutting perfect circles.

And while you can create most box joints at your tablesaw, there are times when you need to take another approach to cut big ones. The clamp-on box-joint jig in **1–4** was designed for use with a handheld router. It's simple to build as well as use.

Of course, these are just four of the router-oriented jigs, fixtures,

and accessories you'll discover in Chapter 2. You may want to build them all!

As you well know, the tablesaw is the workhorse of any woodworking shop. It was with that fact in mind that we assembled the nearly two dozen tablesaw helpers you'll find in the pages of Chapter 3, "Treat Your Tablesaw." A few are as simple and easy to make as the straight-

To cut really large box joints, say for a blanket chest, build this jig described on page 39.

Turn to page 60 to find out how easy it is to make this straight-edge ripping guide.

1–5.

1–4.

edge-ripping jig shown in **1–5**. With it, you'll be able to safely rip straight edges on ragged-edge boards without the telltale nail holes of the alternate method. It takes but a few minutes to assemble, and you'll use it for years. When not in use, store it in a rack with the other stock you have on hand.

For a project that's a bit more challenging, but ever so useful, look on *page 62* for the nifty outfeed table in **1–6**. This handy folding fixture adds only a few inches to your saw when stored, yet provides you with about three feet of stock support beyond the blade. That means safer and more convenient cutting. You can adapt the dimensions to fit any size or style of stationary tablesaw, too. This rip-and-flip extension could be just what the doctor ordered for your shop.

Another of the super-helpful tablesaw aids presented in Chapter 3 is the universal table-saw jig, shown in **1–7**. It has easy-to-adjust

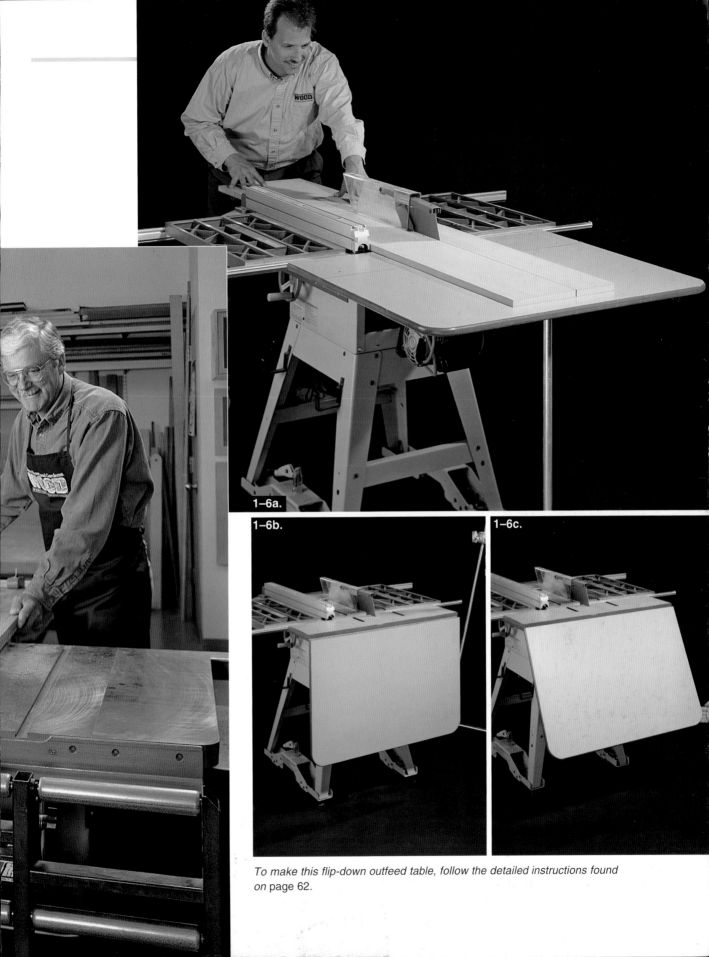

1-6a.

1-6b.

1-6c.

To make this flip-down outfeed table, follow the detailed instructions found on page 62.

hold-downs that ensure safe handling of the workpiece and an infinitely positionable fence. And yes, you can fit it into either the left or right miter-gauge slots for left and right miters.

There's a lot more to this great accessory, too. Turn to *page 66* to find out about it.

On *page 77* you'll discover another easy-to-build and very handy tablesaw fixture. We call it the Texas-size tablesaw fence. With it, as shown in **1–8**, you can stand workpieces, such as raised panels, upright and cut their edges. And it bolts temporarily to your saw's existing fence for safety and accuracy; plus it folds up for storage. We tell you what it takes to make it fit your tablesaw.

Another important woodshop power tool is the bandsaw. It can quickly cut curves and circles, make angled cuts, and resaw thick wood into thinner stock. With help from jigs and accessories, it can do these jobs even more quickly and accurately. In Chapter 4, "Bandsaw Aids," you'll see several projects that will do just that. For example, the resawing jig shown in **1–9** allows you to saw even thin veneer from thicker stock due to its high, flat fence and a feather board that give your workpiece plenty of support.

Top right: *The intersecting mini-channel layout of this jig provides for quick entry and removal of hold-downs. See* page 66.

Bottom right: *Attach this maxi-fence to your tablesaw to cut the edges of large workpieces. Find it on* page 77.

1–7.

1–8.

1–9.

1–10.

Along with the proper blade and tensioning, this jig makes resawing a real pleasure. Turn to page 100.

The circle-cutting attachment for the bandsaw multi-jig lets you saw discs without the traditional center guide hole. Turn to page 107.

If your bandsaw presently is not the precision multi-faceted tool you want it to be, you'll find the solution in the multi-jig on *page 103*. The jig table actually increases the size of your bandsaw table, and the guides steady the blade at tabletop height to minimize blade wander. Its auxiliary circle-cutting guide, shown in **1–10**, allows you to accurately cut disc after disc.

For the ultimate in shop-built bandsaw table systems, turn to *page 109*. As **1–11** shows, this gem of an accessory enables you to crosscut, rip, cut circles, and resaw without adding any difficulty to blade changes.

1–11.

A tapering jig, page 117, is one of the accessories you can build to go with the bandsaw table.

Follow our step-by-step instructions to build it. And while you're at it, you'll want to make the system's three accessories: a tapering jig for table legs, a duplicating jig that produces identical curved parts, and a feather-board/single-point fence for resawing thick as well as narrow stock. With them, you'll have maximum bandsaw versatility.

A drill press does a lot more in a woodworking shop than drill and bore. It can make repetitive holes, even at angles, and sand curved surfaces. In Chapter 5, "Drill-Press Options," you'll learn how to help this reliable power tool do those jobs and more even better. The angle-drilling jig in **1–12** enables you to pivot the table to angles from 0 to 45 degrees, and can even tackle such complicated tasks as compound angles. Our directions and drawings make it easy to build.

For even greater capability, you'll want to check out the drill-press table system on *page 129*. As you can see in **1–13**, this super-versatile accessory has a vertical fence for end-boring and tenoning jobs, adjustable hold-downs, and more!

Of course, not all the projects in Chapter 5 offer challenges. The drill-press pocket-hole jig, shown in **1–14**, will be ready for action after just a few minutes of shop time.

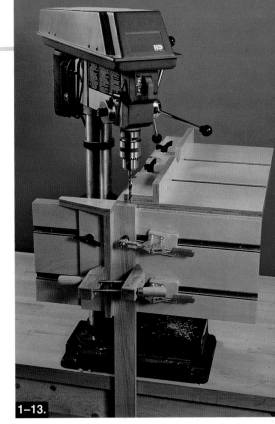

1–13.

Infinite table positioning is only one of this drill-press accessory's great features. See page 129.

The angle-drilling jig on page 124 *gives a standard drill press greater versatility.*

Although you can buy pocket-hole jigs, there is no need to if you make this one, which can be found on page 144.

1–12.

1–14.

1–15.

Take 15 minutes to build this jig and you'll be sanding perfect circles every time. Find it on page 149.

In Chapter 6, "Sanding Solutions," we've gathered together some quite simple, yet highly effective, jigs and fixtures that will earn their place in your shop. If you've always had difficulty sanding circles that are truly round, you'll appreciate the disc-ander circle jig shown in **1–15**.

With it, you'll never again have unwanted center holes in your circles. And how much simpler could it be to build?

Understandably a bit more complex is the sanding table add-on for your tablesaw that you'll see on *page 150*. It's designed for the contractor's saw

we had on hand, but you can alter its dimensions to fit yours. As you can see in **1–16**, this site-specific accessory helps you manage dust from using a hand-held sander, but it's not meant to replace a full-blown dust-collection system.

It's a logical woodworking step to move from sanding to gluing and clamping, so that's why we've separated clamping helpers into their very own chapter. Chapter 7, "Clamping Aids," brings you seven or so devices to help you clamp up your projects, such as the easy-to-make right-angle jig shown in **1–17**. Anytime you're gluing or fastening large workpieces together at a right angle you'll turn to this device. And the more you use this little

Below left: *Note the handy built-in tool tray in this add-on sanding table. Find out how to make it on* page 150.

Below right: *Turn to* page 156 *to see how easy it is to make this simple right-angle clamping jig.*

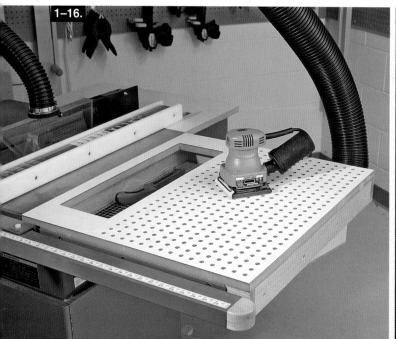

1–16.

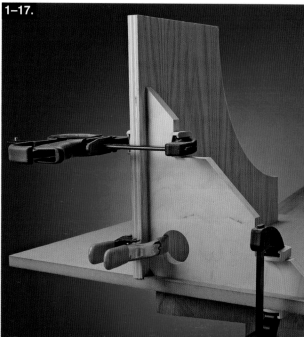

1–17.

1–18.

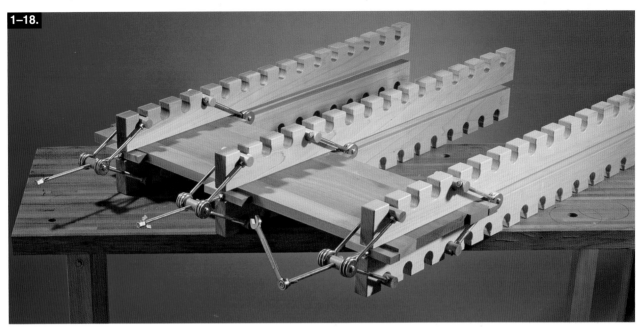

You'll find the patterns and detailed instructions for making these panel clamps on page 160.

helper, the more you'll find for it to do.

Although your workshop probably contains a nice assortment of clamps, we're sure you'll enjoy making and using the panel clamps shown in **1–18**. These pressure-packed clamps end the hassle of edge-joining boards to create a flat panel. Why? Because they'll lay flat on your workbench for one reason. For another, turn to *page 160*. Look over the plans, and then you'll want to make several.

We round out this book with a catchall. Chapter 8, "Assorted Shop Aids," contains an assortment of jigs, fixtures, and accessories that were difficult to categorize, although no less useful than the ones found in the preceding chapters. Take for instance, the multipurpose thickness blocks shown in **1–19**. In the *WOOD®* magazine shop, they're used to

1–19.

On page 167, you'll see how to make and use a set of ever-so-useful hard-maple thickness blocks.

1–20.

This is called the "fail-safe" hinge jig that's perfect for cabinetmaking. See how to make it on page 168.

building the wet-wheel grinder in **1–21**. Used in conjunction with your drill press and a constant water flow, it will put a keen edge on tools without overheating the steel. You can build it for under $100, and we'll show you how to use it, too.

Now that you've finished our content preview, take your time going through the pages of this book and discover the jigs, fixtures, and accessories that we know you'll want to add to your home workshop.

❖

set the position of fences on power tools and come in handy for adjusting the height of saw-blades and router bits. We know you'll find many more uses.

Another handy jig in the last chapter is shown in **1–20**. It faithfully lets you transfer hinge locations from door to carcase or to another door when building cabinets. With it, you'll avoid mistakes when mortising for hinge locations.

Turn to *page 170* to learn how
you can get razor-sharp results with a minimum investment by

Start with a standard horizontal wet-wheel stone to build this grinder for super-sharp edges.

1–21.

CHAPTER 2

Router Accessories

ALTHOUGH ROUTERS FIRST *appeared on the woodworking scene following World War I (the R.L. Carter Company, of Syracuse, New York, sold them in 1919), it's only been in the last 20 years or so that the tool has bloomed in the home workshop. Why? Because of its immense versatility. This lightweight, portable tool, fitted with the appropriate bit, can tackle numerous jobs right out of the box! It'll cut dadoes, rabbets, grooves, slots, decorative edges—you name it. Many, if not most, serious woodworkers have at least two: a handheld model for benchtop work and a table-mounted one for accurate production cutting.*

Despite the router's natural versatility, though, it lends itself well to customizing. That is, with the help of shop-made jigs, fixtures, and accessories, you can double and even triple the tool's capabilities with added accuracy and efficiency. In this chapter, you'll encounter an assortment of router helpers that you easily can build for a fraction of what they'd cost off the shelf (if all of them were even available). So look them over, select what to build, and then get your router really going.

2-1.

Start at the Top

1 Cut both a piece of birch plywood for the panel (A) and a piece of plastic laminate for the skin (B) an inch larger in length and width than the sizes listed in the Materials List.

2 Following the directions on the can, apply contact adhesive to the back of the laminate and the face of the plywood. Bond the laminate to the plywood, holding the laminate about ⅛" back from one edge and one end of the plywood, as shown in "Flush Trimming with a Tablesaw" on the *next page*.

3 Apply pressure with a rubber laminate roller.

4 With the plywood's exposed end and edge in turn against your tablesaw's rip fence, trim about ¼" off the panel's opposite end and edge, cutting through both the plywood and laminate. Now with the just-trimmed end and edge in turn against the fence, cut the panel/skin (A/B) to finished size.

5 Cut the edge bands (C) and the end bands (D) to width, but about 1" longer than the lengths listed. Miter-cut them to fit around the top, as shown on **2–2**. Glue and clamp them in place keeping their top edges flush with the laminate's surface, as shown in "Keep Your Banding Flush and Corners Aligned" on *page 25*.

FULL-SERVICE BENCHTOP ROUTER TABLE

Here's a project (**2–1**) that just belongs in your shop. Here's why:

• You can put it together in a weekend for less than $100 plus the cost of your own wood.

• Its fence adjusts in a flash and locks into T-slotted mini-tracks with the quick twist of two knobs.

• A mini-track built into the fence makes for lightning-fast and solid positioning of homemade feather boards and a bit guard.

• Insert-plate levelers ensure a perfectly aligned tabletop.

• The built-in dust-collection port keeps debris at a minimum.

• It's portable, weighing only 36 pounds (without router), and you can grip the tabletop edges for comfortable carrying.

2–2.

TABLE EXPLODED VIEW

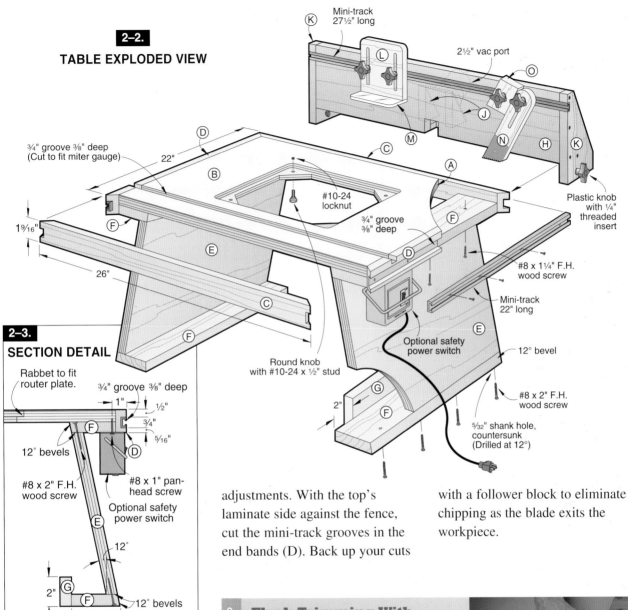

Mini-track
27½" long

K

L

2½" vac port

O

J

M

N

H

K

¾" groove ⅜" deep
(Cut to fit miter gauge)

D

22"

C

B

A

#10-24
locknut

¾" groove
⅜" deep

F

1⁹⁄₁₆"

F

E

D

Plastic knob
with ¼"
threaded
insert

#8 x 1¼" F.H.
wood screw

26"

C

Mini-track
22" long

F

E

Round knob
with #10-24 x ½" stud

Optional safety
power switch

12° bevel

#8 x 2" F.H.
wood screw

G

F

2"

⁵⁄₃₂" shank hole,
countersunk
(Drilled at 12°)

2–3.

SECTION DETAIL

Rabbet to fit
router plate.

¾" groove ⅜" deep

1"

½"

¾"

⁵⁄₁₆"

F

D

12° bevels

#8 x 2" F.H.
wood screw

#8 x 1" pan-
head screw

Optional safety
power switch

E

12°

2"

G

F

12° bevels

3"

6 Install a ¾" dado blade in
your tablesaw, and attach a
tall (about 10") auxiliary fence
to the rip fence. Adjust the blade
and fence to cut the grooves in
the end bands (D) for the mini-
track, where shown on **2–3**.
Test your setup with a piece of
scrap, and make any necessary

adjustments. With the top's
laminate side against the fence,
cut the mini-track grooves in the
end bands (D). Back up your cuts

with a follower block to eliminate
chipping as the blade exits the
workpiece.

Flush Trimming With a Tablesaw

**When applying plastic laminate to a part
like the router-table top's panel, start with
oversize pieces of plywood and laminate.
Apply contact cement to the laminate
and the plywood. Position the laminate
just shy of one edge and end of the
plywood, as shown in 2–4. Run these
edges, free of overhanging laminate,
against your tablesaw fence first when
trimming the top to its finished size.
Cutting both plywood and laminate at the
same time avoids router flush trimming.**

2–4.

Materials List for Full-Service Benchtop Router table

PART	T	W	L	Mtl.	Qty.
	FINISHED SIZE				
TABLE					
A* panel	¾"	20½"	24½"	BP	1
B* skin	1/16"	20½"	24½"	PL	1
C* edge bands	¾"	1 9/16"	26*	M	2
D* end bands	¾"	1 9/16"	22"	M	2
E* legs	¾"	11½"	20½"	BP	2
F* leg cleats	¾"	3"	20½"	M	4
G cord cleat	¾"	2"	16½"	M	1
FENCE					
H* fence	¾"	6"	26 1/32"	M	1
I* fence base	¾"	3"	26 1/32"	M	1
J vac port mounts	¾"	2½""	3⅛"	M	2
K fence brackets	¾"	4¾"	7½"	M	2
GUARD & FEATHER BOARD					
L guard base	¾"	5"	5"	M	1
M guard	¼"	2¾"	5"	A	1
N* feather boards	¾"	1¾"	8"	M	2
O jam blocks	¾"	1¾"	3"	M	2

*Parts initially cut oversize.

Materials Key: BP = birch plywood; PL = plastic laminate; M = maple; A = acrylic.
Supplies: #8 x 1¼" flathead wood screws; #8 x 1½" flathead wood screws; #8 x 2" flathead wood screws; #8 x 1" panhead screws; #8 x 1" brass flathead wood screws (2); ¼" SAE flat washers; contact adhesive; 5-minute epoxy; #10-24 locknuts (4).
Hardware: ¼" hexhead bolts, 1½" long (8); knobs with ¼" threaded inserts (8); miniature knobs with #10-24 x ½" studs (4); 36" mini-track with screws (1); 24" mini-tracks with screws (2); 2½" vac port (1); ⅜ x 12 x 12" acrylic insert plate; Switch: Safety power switch.

Cutting Diagram

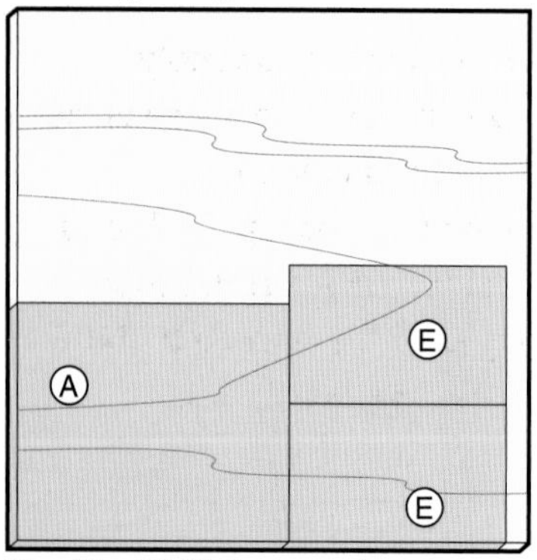
A
E
E

¾ x 48 x 48" Birch plywood

2–5. TABLE TOP

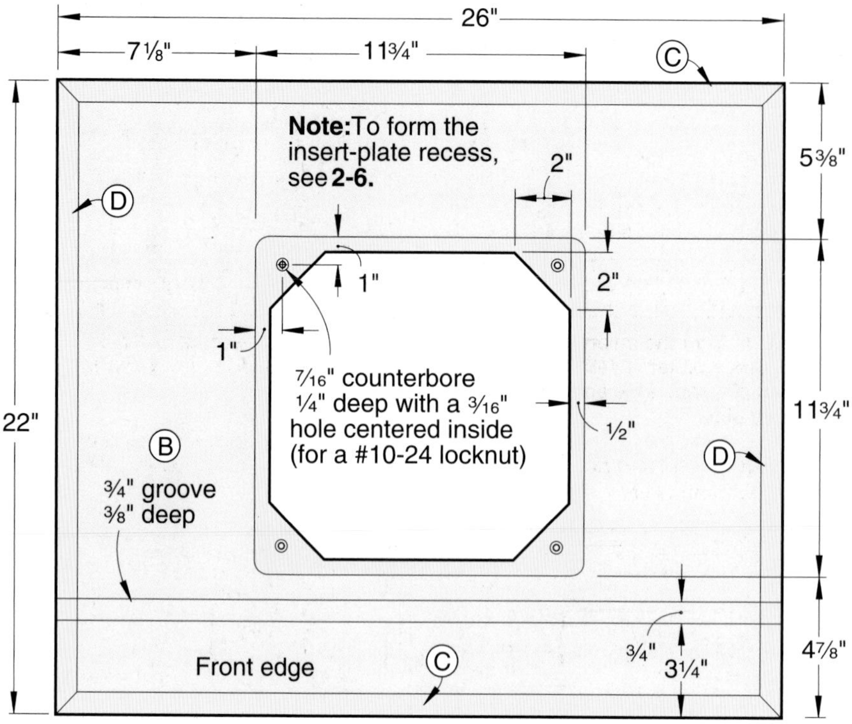
26"
7⅛"
11¾"
C
Note: To form the insert-plate recess, see 2-6.
2"
D
1"
2"
5⅜"
1"
7/16" counterbore ¼" deep with a 3/16" hole centered inside (for a #10-24 locknut)
½"
22"
B
11¾"
D
¾" groove ⅜" deep
¾"
3¼"
4⅞"
Front edge
C

7 With the same dado blade, cut a groove in a piece of scrap, and test the fit of your miter-gauge bar. It should slide freely with very little play. Make any necessary adjustments. With the laminated face down, cut the miter-gauge groove in the top, where shown on **2–5**. Back up the cut with a follower block to eliminate chipping.

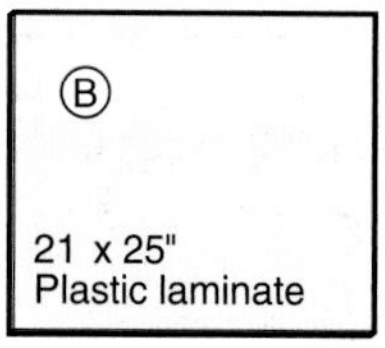
B
21 x 25"
Plastic laminate

Fit the Insert Plate and Install Plate Levelers

1 Follow the eight steps in **2–6** to create the router-table top's insert-plate recess.

2 With the insert-plate recess formed, drill 7/16" counterbores ¼" deep in each corner for #10–24 locknuts, where shown on **2–5**. Make sure that, when placed in the counterbores, the locknuts are flush with the surface of the recess. Drill 3/16" holes through the centers of the counterbores.

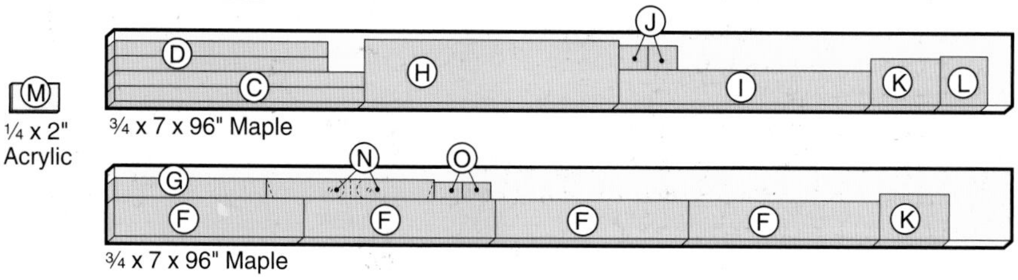
M
¼ x 2"
Acrylic
J
D
C
H
I
K
L
¾ x 7 x 96" Maple
N
O
G
F
F
F
F
K
¾ x 7 x 96" Maple

2–6. FORMING THE INSERT-PLATE RECESS

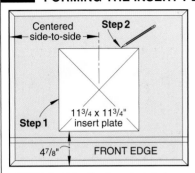

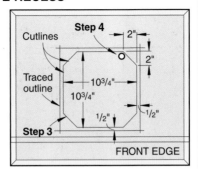

STEP 1 Trim the insert plate to size, and position it 4⅞" from the tabletop's front edge and centered side-to-side.

STEP 2 Trace the outline of the plate onto the tabletop.

STEP 3 Lay out and mark the opening cutlines inside the traced outline.

STEP 4 Drill a blade start hole, and use your jigsaw to cut the opening.

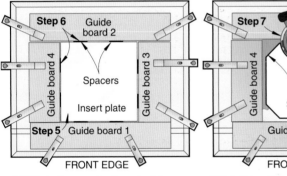

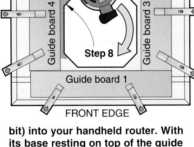

STEP 5 Secure the insert plate inside the traced outline with double-faced tape.

STEP 6 Clamp the guide boards around the plate, spacing each board away from the plate with business-card shims.

STEP 7 Remove the insert plate and shims. Chuck a straight bit with a top-mounted pilot bearing (pattern

bit) into your handheld router. With its base resting on top of the guide boards, adjust the router to cut ⅛" into the tabletop.

STEP 8 Guiding the bit's pilot bearing along the guide board's inside edges, begin routing the recess. Make additional passes, lowering the bit each time until you reach a depth of about ¹⁄₃₂" greater than the thickness of the insert plate.

SHOP TIP

Keep Your Banding Flush and Corners Aligned

Make alignment blocks by cutting 2 x 2" notches out of four 4 x 4" pieces of ¾" plywood. (The notches let you see the mitered corners.) Clamp them to the top, as shown in 2–7. Use scrap blocks underneath the top to space the clamps away from the banding. Now, glue and clamp the banding to the top, keeping it tight against the alignment blocks.

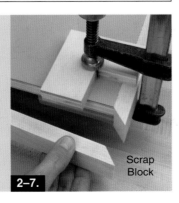

Scrap Block

2–7.

3 Referring to "Add Insert-Plate Levelers to Your Table" on *page 27*, epoxy locknuts into the counterbores. Finish-sand the bands (C, D) to 220 grit. Ease the sharp laminate edges of the miter-gauge slot and insert-plate recess with a cabinet scraper.

Build a Sturdy Base

1 Cut the legs (E) and leg cleats (F) to length, but about 1" wider than listed. Tilt your table-saw blade 12°, and bevel-rip the edges of the legs and leg cleats, where shown on **2–3**. Cut the cord cleat (G) to size.

2 Glue and clamp the leg cleats (F) to the legs (E). Then drill pilot and countersunk shank holes through the leg cleats into the legs. Drive in the screws, and remove the clamps. Glue and clamp the cord cleat to the leg cleat. Finish-sand the leg assemblies to 220 grit.

Note: When storing the router table, coil the router and switch cords and stow them under the table, wedging them between the leg and the cord cleat.

3 Place the top assembly upside down on your bench. Glue and clamp the leg assemblies to the top. Drill pilot and countersunk shank holes through the leg cleats into the top. Drive in the screws.

Make an Accurate Fence

1 Forming straight, square edges on your fence parts is essential for making a straight fence. Start by cutting the fence (H) and the fence base (I) 1/2" wider and 1" longer than the sizes listed. Joint one edge of each board. Next set the fence on your tablesaw 1/32" over the finished width, and rip the parts. Set your jointer's depth to 1/32" and joint the freshly cut edge. Check the length of your tabletop and add 1/32" to this measurement. Cut the fence and fence base to this length. (The added 1/32" allows the fence to slide easily.) Bandsaw centered 1 1/2 x 1 1/2" bit-clearance notches in both parts. Glue and clamp the fence and fence base together.

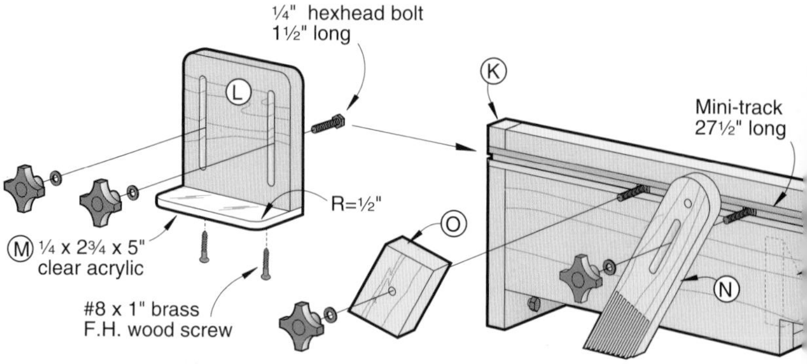
1/4" hexhead bolt 1 1/2" long
L
K
Mini-track 27 1/2" long
R=1/2"
O
M 1/4 x 2 3/4 x 5" clear acrylic
#8 x 1" brass F.H. wood screw
N

2-8. FENCE EXPLODED VIEW

2 Cut two 3/4 x 4 3/4 x 7 1/2" blanks for the fence brackets (K). Fasten the two blanks together with double-faced tape. Mark the diagonal cut and the location of the 1/4" hole on the top blank, where shown on **2–9**. Bandsaw and sand to the marked line, and drill the hole. Separate the brackets.

3 Glue and clamp the fence brackets (K) to the fence (H/I), making sure the brackets' edges are flush with the fence's face. Drill pilot and countersunk shank holes through the brackets into the fence, where shown, and drive in the screws. With your dado blade adjusted to the width of the mini-track, cut the dado in the fence (K/H/K), where shown on **2–8**. Finish-sand the fence assembly to 220 grit.

4 Cut the vac port mounts (J) to the size and shape shown on **2–12**. Dry-position the mounts and check their placement with

2-9. FENCE BRACKETS

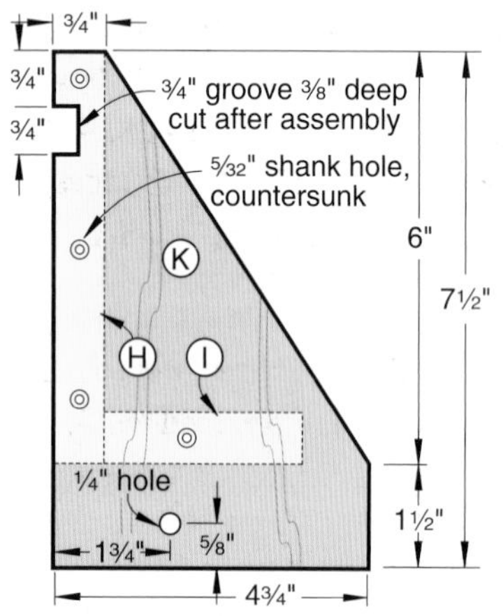
3/4"
3/4"
3/4"
3/4" groove 3/8" deep cut after assembly
5/32" shank hole, countersunk
K
6"
7 1/2"
H
I
1/4" hole
1 3/4"
5/8"
1 1/2"
4 3/4"

2-10. GUARD

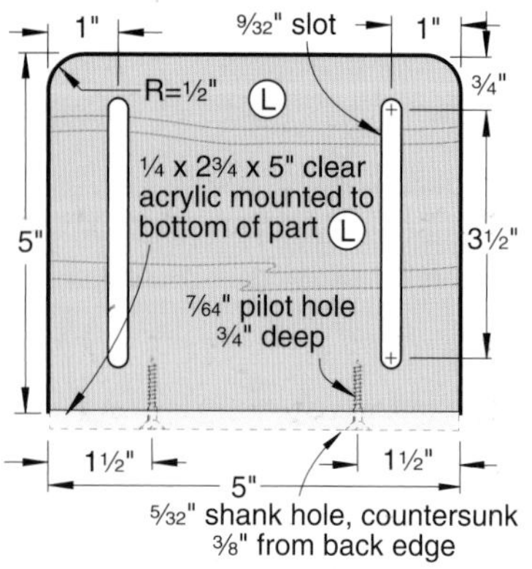
1"
9/32" slot
1"
R=1/2"
L
3/4"
1/4 x 2 3/4 x 5" clear acrylic mounted to bottom of part L
5"
3 1/2"
7/64" pilot hole 3/4" deep
1 1/2"
1 1/2"
5"
5/32" shank hole, countersunk 3/8" from back edge

SHOP TIP

Make a Self-Gauging Feather Board

Here's a quick way to set your feather board to apply the proper pressure. Trim the first feather 1/8" shorter than the others, where shown below. When you use your feather board, place this short gauging feather on top of your workpiece. Now, keeping the other feathers parallel to the router-table top, tighten the mounting knob.

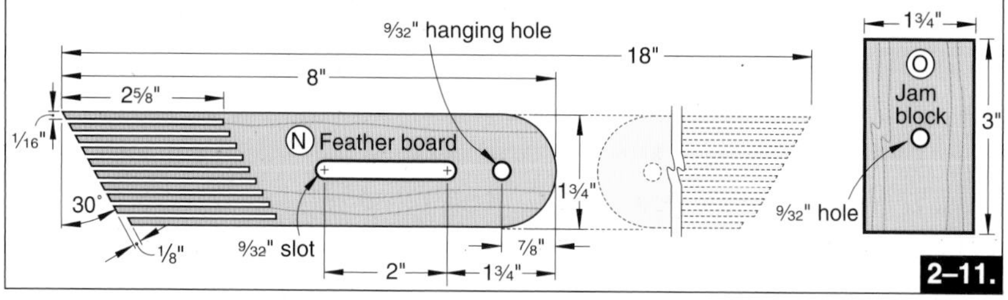
9/32" hanging hole
18"
8"
2 5/8"
1/16"
N Feather board
30°
1/8"
9/32" slot
2"
1 3/4"
7/8"
1 3/4"
1 3/4"
O Jam block
3"
9/32" hole

2-11.

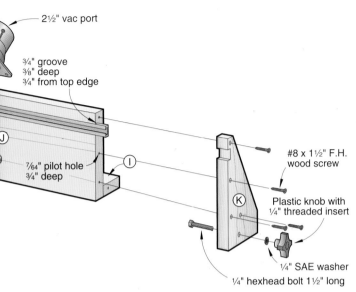

2½" vac port

¾" groove
⅜" deep
¾" from top edge

J

⁷⁄₆₄" pilot hole
¾" deep

I

#8 x 1½" F.H.
wood screw

K

Plastic knob with
¼" threaded insert

¼" SAE washer
¼" hexhead bolt 1½" long

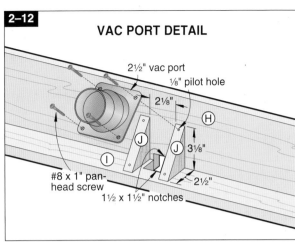

2–12 **VAC PORT DETAIL**

2½" vac port
⅛" pilot hole
2⅛"
H
J
3⅛"
I
#8 x 1" pan-
head screw
2½"
J
1½ x 1½" notches

your vac port. Glue and
clamp the mounts in place.
With the glue dry, use the port
to mark the mounting screw
locations. Drill screw pilot holes
and set the vac port aside.

Now, Make the Guard

1 Cut the guard base (L) to size.
Sand the ½" radii on the top
corners, where shown on **2–10**.
To form the mounting slots, drill
⁹⁄₃₂" holes where shown, draw
lines from hole to hole, and scroll-
saw along the lines. Finish-sand
the base to 220 grit.

2 Cut ¼" acrylic to size for the
guard (M). Disc-sand ½" radii
on the outside corners, where
shown on **2–8**. Adhere the guard
to the base with double-faced
tape, keeping the back edges
flush. Drill pilot and countersunk
shank holes through the guard
(M) into the base (L). Remove the
guard, and set it aside.

Make the Feather Boards

1 Select a straight-grained piece
of ¾"-thick maple, and cut a
¾ x 2 x 18" blank for the feather
boards. Using your tablesaw and
miter gauge, trim 30° angles on
both ends of the blank, where
shown on **2–11**. Mark angled lines
across the blank's width, 2⅝"
from each end, and then mark the
feather boards' radius ends.

2 Install a regular (⅛" wide)
blade in your tablesaw and
raise it 2" high. Set the rip fence
¹⁄₁₆" from the blade. With the long
edge of the blank against the fence,
cut in to the marked line, and then
carefully pull the board straight
back from the blade. A padded
jointer pushblock works well for
this operation. Flip the board end
for end and repeat. Reset the
fence at ¼" and repeat the cut on
each end. Repeat cutting the

feathers at ³⁄₁₆" intervals up to
1¾". With the fence set at 1¾",
lower the blade to 1", and cut the
blank to finished width.

3 Drill the ⁹⁄₃₂" hanging and
slot-starting holes in the
feather boards (N), where shown
on **2–11**. Mark and scroll-saw the
slots, and bandsaw the rounded
ends. Finish-sand the feather
boards to 220 grit.

4 Cut the jam blocks (O) to size
and drill the centered ⁹⁄₃₂" holes.
Finish-sand them to 220 grit.

*Note: The jam blocks are
positioned against the feather
boards to prevent them from
pivoting when applying pressure
to a workpiece.*

Add Insert-Plate Levelers to Your Table

Adjusting your router table's
insert plate perfectly flush with
the top is as easy as installing
locknuts in the corners of the
plate's recess. Once you've
drilled counterbored holes to

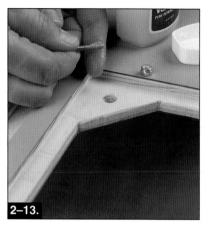

2–13.

To keep the epoxy from sticking to it, apply petroleum jelly to a #10–24 x 2" machine screw.

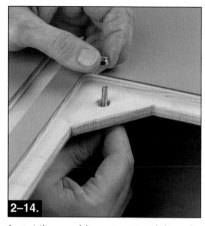

2–14.

Insert the machine screw, and thread on a stop nut until the end of the screw is flush with the top of the nut.

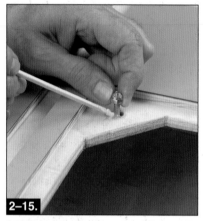

2–15.

Mix some 5-minute epoxy. Fill the counterbore with epoxy while pulling up on the locknut.

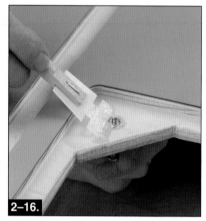

2–16.

Pull down on the machine screw to seat the nut in the counterbore. Scrape off the excess epoxy.

accept #10–24 nuts in all four corners, here's how to proceed (**2–13** to **2–16**). When the epoxy hardens, replace the 2"-long machine screws with a ½"-long threaded stud.

Apply a Finish, and Install the Hardware

1 Touch up the finish sanding where needed. Apply two coats of a penetrating oil finish to all the wood parts, including the miter-gauge slot and the insert-plate recess, following the instructions on the can. An oil finish is easier to reapply after the finish is worn than paint or varnish. It also seals the miter-gauge slot and insert-plate recess without building up and interfering with the fit.

2 Hacksaw mini-track to the lengths of the table ends and fence. You'll have to drill and countersink new mounting holes at the cut ends. Using the holes in the mini-track as guides, drill pilot holes into the table and fence, and screw the track in place.

3 Mount the optional switch, where shown on **2–2** and **2–3**.

4 Screw the guard (M) to the guard base (L) with #8 1" brass flathead wood screws. Attach the assembled guard, feather boards, and jam blocks to the fence and the fence to the table with hexhead bolts, washers, and knobs, as shown. Screw the vac port to the mounts.

5 Screw the insert-plate leveling knobs into the locknuts. Sand the insert plate's corners to match the corners of the insert recess.

❖

SLIDING TOP FOR ROUTER TABLES

Routing dadoes—grooves across the grain—poses several challenges, especially on narrow stock. Securing the workpiece, spacing the dadoes, and guiding the router straight over the stock become even more difficult on small workpieces. This sliding top for your router table (**2–17**), from *WOOD*® magazine reader C. E. Rannefeld, of Decatur, Alabama, makes dado-routing easy. Start with a piece of ⅛"-thick tempered hardboard as wide as the front-to-back dimension of your router table and about 4" longer than the end-to-end distance. Attach a 1 x 2"

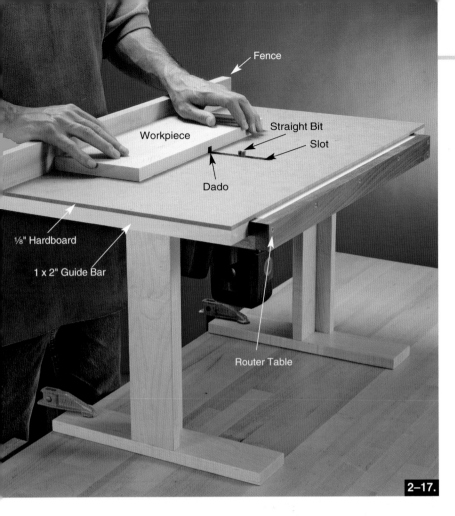

Fence

Straight Bit

Workpiece

Slot

Dado

1/8" Hardboard

1 x 2" Guide Bar

Router Table

2–17.

guide bar across each end on the underside of the hardboard, locating them so the hardboard slides without excessive side play. Chuck the straight bit for dadoing into the router. Raise the bit the distance above the hardboard equal to the depth you want the dado to be. Push the hardboard sliding tabletop into the bit, and cut a slot about halfway across the hardboard. Notch a 1 x 2" fence to clear the bit, and mount it at the back of the sliding top. To rout dadoes, hold the work-piece firmly against the fence and slowly slide the tabletop across the bit. Add a stopblock for repetitive cuts.

DO-IT-ALL ROUTER-TABLE FENCE

Is an ordinary fence limiting the performance of your table-mounted router? Now you can move your routing into the big leagues with this feature-packed upgrade (**2–18** to **2–23**). To add flexibility, the fence is designed to attach to a router table in just about every imaginable way. Use threaded inserts, T-track, or simply clamp the ends to your router table. If the fence is for a router table built into your tablesaw extension, an optional cleat enables you to clamp it to the tablesaw rip fence.

Start with the Fence Body

1 From 1/2" plywood (we used Baltic birch), cut the upright (A) and base (B) to the sizes listed on the Materials List on *page 37*. Adjust a dado blade to the thickness of your 1/2" plywood, and cut 1/8"-deep dadoes across the widths of the parts, where shown on **2–24**.

Note: If your router table already has threaded inserts or T-track for mounting and securing a fence, make sure the location of the braces does not interfere with it.

Because these dadoes house the braces (C) [see **2–25**, on *page 31*], they must align perfectly. Position the tablesaw rip fence as a stop 4" from the blade and, using the miter gauge to steady the parts, cut all four of the outside dadoes. Reposition the fence 11" from the blade and cut the four inside dadoes. Now cut the 1/2" rabbet along the bottom edge of the upright.

Note: For the upright to be square to the base after assembly, the dadoes and rabbet must be uniform in depth. Make two passes over the blade to make certain the bottoms of your cuts are completely cleaned out.

2 Lay out the centers of the 5/16" holes that form the ends of the slots in the upright (A), where shown on **2–24**. For the movable face parts F and G to work properly, the slots must be perfectly aligned, so use your drill

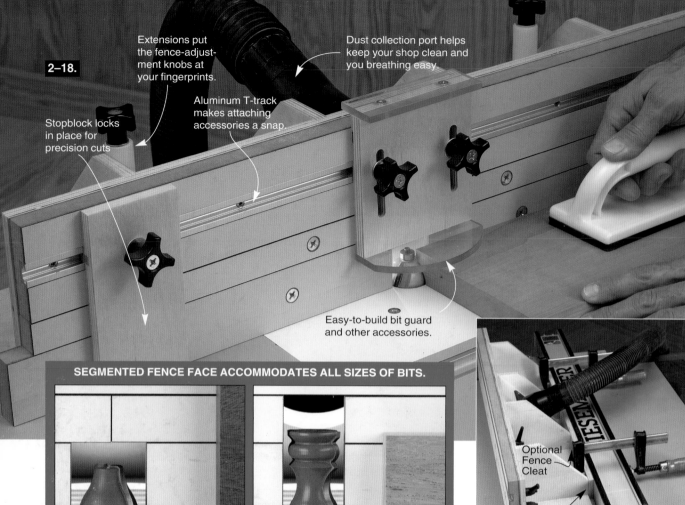

2–18.

Extensions put the fence-adjustment knobs at your fingerprints.

Dust collection port helps keep your shop clean and you breathing easy.

Aluminum T-track makes attaching accessories a snap.

Stopblock locks in place for precision cuts

Easy-to-build bit guard and other accessories.

SEGMENTED FENCE FACE ACCOMMODATES ALL SIZES OF BITS.

2–19.

The 2"-high lower portion of the fence opens to house the majority of your router bits.

2–20.

To accommodate tall bits, such as this crown molding cutter, open the 1"-tall center portion.

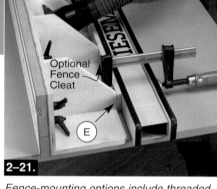

Optional Fence Cleat

E

2–21.

Fence-mounting options include threaded inserts, T-track, and clamps.
For a saw-table-mounted router, clamp it to the rip fence.

press and its fence to align the bit and drill the holes. If you plan to secure the fence to your router table with threaded inserts, drill slot-end holes in the base (B), where shown. To locate base slots for a router-table top that already has threaded inserts, measure the center-to-center distance and center this dimension on the base. Now scrollsaw the slots, as shown in **2–27**, on *page 32.*

If you plan to install T-track in your router-table top, drill only

2–22.

A one-piece feather board safely and firmly holds down workpieces for consistent profiles.

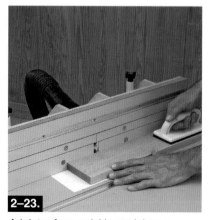

2–23.

A jointer face quickly straightens edges or removes saw marks for edge-gluing.

2–24. ■ **FENCE BODY PARTS**

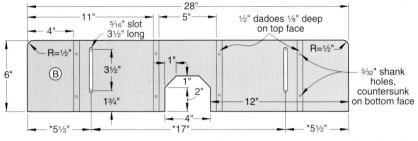

FENCE BASE
(Top face shown)

Note: *If your router table already has threaded inserts or T-track, space the slots or holes to match. See the instructions.

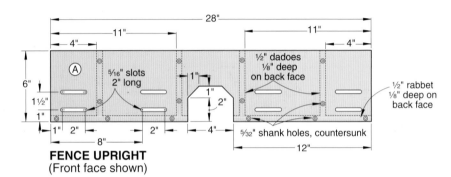

FENCE UPRIGHT
(Front face shown)

3 Lay out the bit clearance cutouts in the upright (A) and base (B), where shown on **2–24**, and scroll-saw or jigsaw them to shape. Then sand ½"-radii on the rear corners of the base.

4 Using your drill press, drill countersunk holes in the upright (A) and base (B) centered on the dadoes and rabbet, where shown on **2–24**. Finish-sand the parts. Then glue and clamp the upright and base together, keeping the ends flush. Using the holes in the upright as guides, drill pilot holes into the base, and drive the screws.

5 Cut the braces (C) to size, and then cut them to the shape shown on **2–25**. Finish-sand the braces. Now clamp them in the

the slot-end holes closest to the front edge of the base (B). To locate base holes for a router-table top that already has T-track, measure the center-to-center distance and center this dimension on the base. If you plan to clamp the fence to your router table, no slots or holes are needed in the base.

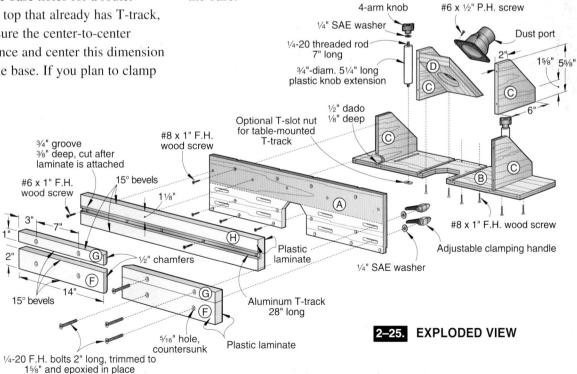

2–25. ■ **EXPLODED VIEW**

upright (A) and base (B) dadoes, making sure they fully seat in each part. Using the holes in the upright and base as guides, drill pilot holes into the base, and drive the screws.

Note: For the upright to be square to the base, the front edge of each brace must be square to its bottom edge.

6 Check the distance between the center braces (C), and cut the dust port panel (D) to size. Then cut 45° bevels on the ends, where shown on **2–26**. Center and draw a 3"-diameter circle on the panel with a compass, drill a blade start hole inside the circle, and saw out the hole. Finish-sand the panel. Now apply glue to the bevels and clamp the panel in place between the braces, with its top edge flush with the top edge of the upright (A).

7 If you will clamp the fence to your tablesaw rip fence for use with an extension-table mounted router, as shown in

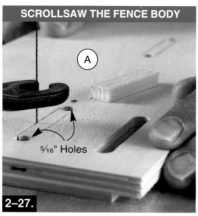

SCROLLSAW THE FENCE BODY

A

⁵⁄₁₆" Holes

2–27.

With the slot-end holes drilled, draw tangent lines connecting each pair of holes, and scrollsaw the slots.

2–28, measure the height of the rip fence, cut the optional fence cleat (E) to size, and finish-sand it. Clamp the cleat to the braces (C), where shown on **2–29**. Drill countersunk holes through the cleat and into the braces, and drive the screws.

Make the Segmented Face

1 For the fence faces (F, G, H), cut two pieces of plastic

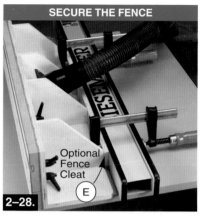

SECURE THE FENCE

Optional Fence Cleat

E

2–28.

For a tablesaw-extension mounted router, adding the optional cleat (E) allows you to clamp the fence to the tablesaw rip fence.

laminate and a piece of ¾" medium-density fiberboard (MDF) to 7 x 29". (We used Formica brand laminate in no. 464 Graystone color.) Adhere the laminate to both sides of the MDF with contact adhesive. True one edge and one end of the laminated blank on your tablesaw. Then cut the lower face (F), center face (G), and upper face (H) to size. Using a 15° bevel laminate trimming router bit, bevel the ends and edges of the parts.

2–26. **DUST PORT PANEL**

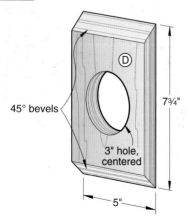

45° bevels

7¾"

3" hole, centered

5"

D

2–29. **OPTIONAL TABLESAW FENCE CLEAT**

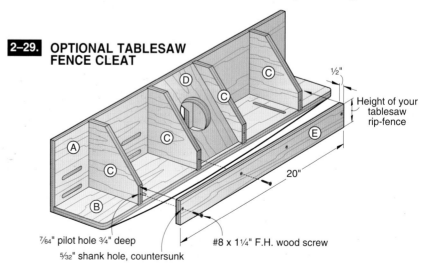

½"

Height of your tablesaw rip-fence

D

C

C

C

A

C

E

20"

B

⁷⁄₆₄" pilot hole ¾" deep
⁵⁄₃₂" shank hole, countersunk

#8 x 1¼" F.H. wood screw

2 Cut ½" chamfers on the inside ends of the lower face (F) and center face (G), where shown on **2–25**. Then drill ⁵⁄₁₆" holes in the parts. Countersink the holes so the head of a ¼" flathead bolt is slightly below the laminate surface.

Note: The holes are oversize to allow room for epoxy when permanently mounting the bolts in the faces (F, G).

3 Install a ¾" dado blade in your tablesaw and cut a ⅜"- deep groove for the aluminum T-track in the upper face (H). Fit the track in the groove, ends flush with the upper face ends. Using the pre-drilled holes in the track as guides, drill shank holes through the upper face.

Apply a Finish and Assemble the Fence

1 Cover the plastic laminate surfaces with masking tape. Then apply a clear finish to all the parts. (To adequately seal the MDF edges of the fence faces, we brushed on four coats of satin polyurethane, sanding with 220-grit sandpaper between coats. We finished the fence body with two coats of aerosol satin polyurethane, sanding between coats.) Remove the masking tape.

2 Cut 2"-long flathead bolts to 1⅝", as indicated on **2–25**. To protect the plastic laminate from excess epoxy, cover the holes in the faces (F, G) with plastic packing tape. Cut around

the countersinks with a utility knife. Epoxy the bolts in the holes, as shown in **2–31**. When the epoxy cures, remove the tape. Use a chisel to pare away any excess epoxy that protrudes beyond the plastic laminate surface.

3 Clamp the fence body to a flat surface. Insert the lower face (F) and center face (G) bolts in the slots in the upright (A). Insert business or playing cards between them as spacers, as shown in **2–32**, and secure the faces with washers and adjustable clamping handles, as shown on **2–25**. Now position the upper face (H) on the center face, and insert card spacers between them. Make sure the ends of the upper face and upright are flush, and clamp the face to the upright. Fasten the upper face to the upright, as shown in **2–32**.

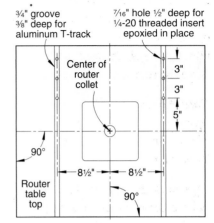

2–30. **INSTALLING T-TRACK OR THREADED INSERTS**

¾" groove ⅜" deep for aluminum T-track

⁷⁄₁₆" hole ½" deep for ¼-20 threaded insert epoxied in place

Center of router collet

Router table top

90°

90°

3"
3"
5"

8½" — 8½"

4 Position the dust port over the hole in the dust port panel (D). Using the holes in the port as guides, drill pilot holes into the panel and screw the dust port snugly in place.

5 If needed, install aluminum T-track or ¼-20 threaded inserts in the grooves in your router table top, where shown on **2–30**.

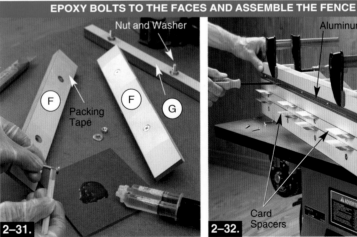

EPOXY BOLTS TO THE FACES AND ASSEMBLE THE FENCE

Nut and Washer

Aluminum T-Track

F
Packing Tape
F
G

Card Spacers

2–31.

Apply epoxy to the bolt shanks, and insert them into the holes. Snug the bolts in place with washers and nuts, making sure they are perpendicular to the surface. When the epoxy cures, remove the nuts and washers.

2–32.

With card spacers between the fence faces (F, G, H), install the T-track in the upper face (H). Using the holes in the track and face as guides, drill pilot holes into the upright (A), and drive the screws.

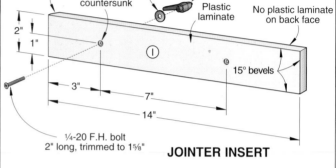

holes in the infeed face. Mask off the laminate and apply polyurethane. Then trim two 2"-long flathead bolts to 1⅝" and epoxy them in place.

2–34.

1/4" SAE washer

Adjustable clamping handle

1/2" chamfer

5/16" hole, countersunk

Plastic laminate

No plastic laminate on back face

2"

1"

15° bevels

3"

7"

14"

1/4-20 F.H. bolt 2" long, trimmed to 1⅝"

JOINTER INSERT

6 To clamp the fence to a router table equipped with T-track or threaded inserts, cut two pieces of ¼–20 threaded rod 7" long. For T-track, thread the rods into the raised-collar side of the T-slot nuts so ¹⁄₁₆" protrudes from the bottom of the nut. Then slide the nuts into the tracks, and drop the fence down over the rods. Slip the plastic knob extensions onto the rod and add washers and four-arm knobs, as shown on **2–25**. Tighten the knobs enough to secure the fence. Now fill the knob recesses with epoxy, fixing the threaded rods in place. For threaded inserts, thread the rods ⅜" into the inserts. As you did with the T-track, add the fence, knob extensions, washers, and knobs. Tighten the knobs and add epoxy.

Four Super-Handy Accessories for Your New Fence

1. Jointer Insert (2–33) Helps You Straighten Edges

Make the Infeed Face

From ¾" MDF, cut the infeed face (I) to size. (In use, the infeed face replaces the right-hand lower face.) Cut a 2½ x 14½" piece of plastic laminate and adhere it to one side of the MDF with contact adhesive. Using a 15° laminate bevel trimming bit, trim the excess laminate. Cut a ½" chamfer on one end of the infeed face, as shown on **2–34**.

As you did when making the lower faces, drill ⁵⁄₁₆" countersunk

Joint with Your Fence

To joint an edge on your router table, remove the right-hand lower fence face (F) and replace it with the infeed face (I). Then chuck a straight router bit in the router and align the left-hand lower fence face (F) with the bit, as shown in **2–35**. Now slide the infeed face (I) to within ⅛" of the bit and secure it with washers and adjustable clamping handles. Make test cuts and fine-tune the fence position by loosening

JOINT WITH YOUR ROUTER TABLE FENCE.

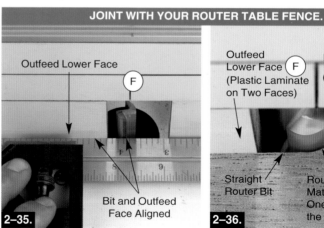

Outfeed Lower Face

F

Bit and Outfeed Face Aligned

2–35.

Position the left-hand (outfeed) lower fence face (F) ⅛" from the cutting edge of the bit. Using a straightedge and moving the fence, align the fence face with the bit. Then clamp the fence in place.

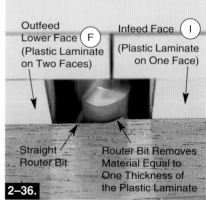

Outfeed Lower Face F (Plastic Laminate on Two Faces)

Infeed Face I (Plastic Laminate on One Face)

Straight Router Bit

Router Bit Removes Material Equal to One Thickness of the Plastic Laminate

2–36.

Slide your stock along the infeed face and into the bit. Because the jointer has plastic laminate on only one side, the bit removes stock equal to the thickness of the laminate.

one end and lightly tapping it forward or backward with a mallet to precisely align the bit with the outfeed face. Joint your stock, as shown in **2–36**.

2. Adjustable Bit Guard (2–37) Protects Fingers and Deflects Chips

Cut and Assemble the Parts

Cut the guard body to size. As you did when slotting the fence upright (A), drill ⁵⁄₁₆" holes, as shown on **2–38**. Then connect the holes with tangent lines and saw out the slots. Finish-sand the body and apply a clear finish.

From ¼" clear acrylic, cut a 1½ x 4½" piece for the handle and a 2¼ x 4¼" piece for the shield. Make a copy of the shield pattern, *below*, and adhere it to the shield blank with spray adhesive. Bandsaw and sand the curve. Then drill countersunk holes in the handle and shield, where shown. (The holes are oversize to prevent cracking the acrylic.) Sand the edges of both parts smooth.

Centering the handle on the body, and aligning the straight edge of the shield flush with the back face of the body, use the holes as guides and drill pilot holes. Remove the masking sheet from the acrylic, and screw the parts to the body.

Thread two 1¾"-long flathead bolts into two 4-arm knobs, leaving the bolt heads protruding ½" from the top of each knob. Then apply epoxy under the bolt heads and drive them the rest of the way.

With the epoxy cured, slip washers on the bolts, insert them in the slots, and thread on T-slot nuts with the raised collars toward the knob.

Keep Your Fingers Safe

To use the bit guard, slide the T-slot nuts into the T-track and center the shield over the bit. Grasp the guard by the ¼" acrylic handle, adjust its height to clear the stock you will be routing, and tighten the knobs.

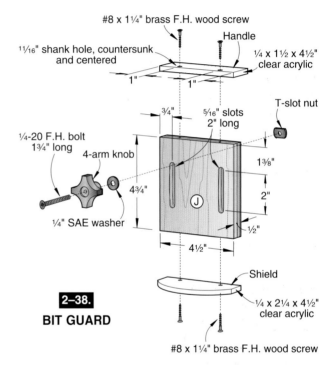

#8 x 1¼" brass F.H. wood screw

Handle

¹¹⁄₁₆" shank hole, countersunk and centered

¼ x 1½ x 4½" clear acrylic

1"   1"

¾"   ⁵⁄₁₆" slots 2" long

T-slot nut

¼-20 F.H. bolt 1¾" long

4-arm knob

1³⁄₈"

4¾"

2"

J

¼" SAE washer

½"

4½"

2–38.
BIT GUARD

Shield

¼ x 2¼ x 4½" clear acrylic

#8 x 1¼" brass F.H. wood screw

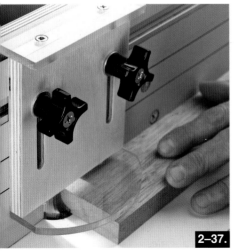

2–37.

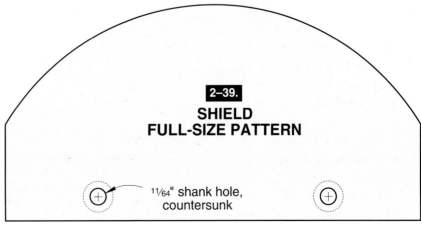

2–39.
SHIELD
FULL-SIZE PATTERN

¹¹⁄₆₄" shank hole, countersunk

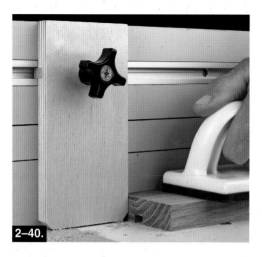

2–40.

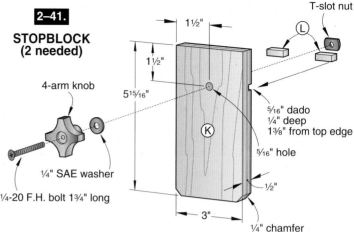

2–41.

**STOPBLOCK
(2 needed)**

4-arm knob

¼" SAE washer

¼-20 F.H. bolt 1¾" long

T-slot nut

L

K

1½"

1½"

5¹⁵⁄₁₆"

⁵⁄₁₆" dado
¼" deep
1⅜" from top edge

⁵⁄₁₆" hole

½"

3"

¼" chamfer

3. Locking Stopblocks (2–40) Enable Precise Stopped Cuts

Make the Bodies and Cleats

Cut two stop bodies (K) to size. Then cut a ⁵⁄₁₆" dado ¼" deep in the back of each one, where shown on **2–41.** Drill a ⁵⁄₁₆" hole centered in the dadoes and on the width of the bodies. Now cut ¼" sawdust-relief chamfers on the bottom corners. Finish-sand the bodies.

Resaw and plane a ⁵⁄₁₆ × ½ × 10" blank for the cleats (L), checking its fit in the stop body dadoes. Cut the cleats to length, and glue and clamp them in the dadoes with the ends flush with the edges of the stop body. Apply a clear finish.

Epoxy 1¾" flathead bolts into two, 4-arm knobs. Install the knobs and washers in the stopblocks and add T-slot nuts.

Make Stopped Cuts

To use the stopblocks, slide the cleats (L) and T-slot nuts into the T-track. Using a ruler, position the stopblocks the required distance from the bit and tighten the knobs.

4. Feather Board Holds Pieces for Consistent Cuts

Machine the MDF Blank

Cut a piece of ¾" medium-density fiberboard to the size listed. Enlarge a copy of the feather board pattern on *page 37*, and adhere it to the blank with spray adhesive. Install a blade in your bandsaw that cuts a ¹⁄₁₆" kerf. (We used a ½" resawing blade.) Cut the feathers, as shown in **2–44.**

Chuck a ⁵⁄₁₆" bit in your drill press and drill the pivot

hole and the holes at the ends of the curved slot. Then scroll-saw the slot. Now bandsaw the curved edge of the feather board. Apply a clear finish. Epoxy 2" flathead bolts into two 4-arm knobs, as directed in the instructions for the bit guard on *page 35*. Install the knobs and add T-slot nuts, where shown on **2–43.**

How to Use the Feather Board

Mount the feather board on the fence by sliding the T-slot nuts into the T-track, positioning the pivot hole on the right-hand (infeed) side of the bit and the curved slot on the left-hand (outfeed) side. Center the feather

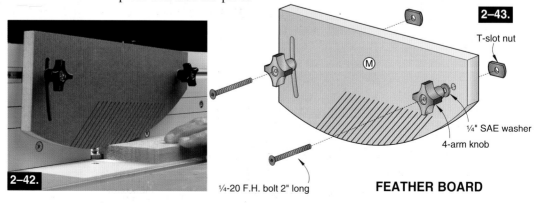

2–42.

2–43.

T-slot nut

M

¼" SAE washer

4-arm knob

¼-20 F.H. bolt 2" long

FEATHER BOARD

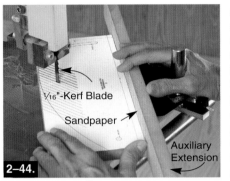

1/16"-Kerf Blade

Sandpaper

Auxiliary Extension

2–44.

Attach an auxiliary extension to your miter gauge, and adhere sandpaper to the extension to keep the workpiece from slipping. Adjust the miter gauge to 45°, and saw the feathers along the pattern lines.

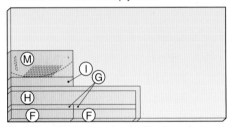

Cutting Diagram

1/2 x 24 x 30" Baltic birch plywood

3/4 x 3/4 x 10" Maple
*Plane or resaw to the thickness listed in the Materials List.

3/4 x 24 x 48" Medium-density fiberboard

Materials List for Do-It-All Router-Table Fence

PART	FINISHED SIZE			MTL.	QTY.
	T	W	L		
FENCE					
A upright	1/2"	6"	28"	BP	1
B base	1/2"	6"	28"	BP	1
C braces	1/2"	5 5/8"	6"	BP	4
D dust port	1/2"	5"	7 3/4"	BP	1
E optional fence cleat	1/2"	†	20"	BP	1
F* lower faces	3/4"	2"	14"	MDF	2
G* center faces	3/4"	1"	14"	MDF	2
H* upper face	3/4"	3	28"	MDF	1
JOINTER					
I infeed face	3/4"	2"	14"	MDF	1
BIT GUARD					
J guard body	1/2"	4 1/2"	4 3/4"	BP	1
STOPBLOCKS					
K stop bodies	1/2"	3"	5 5/16"	BP	2
L* cleats	5/16"	1/2"	3/4"	M	4
FEATHER BOARD					
M feather board	3/4"	5 3/4"	14"	MDF	1

†Height of your tablesaw rip fence. See the instructions.
*Parts initially cut oversize. See the instructions.

Materials Key: BP = Baltic birch plywood; MDF = medium-density fiberboard; M = maple.
Supplies: Contact cement; epoxy; spray adhesive.
Blades and Bits: Stack dado set; 15° bevel laminate-trimming router bit.
Hardware: 24 x 30" plastic laminate; 1/4" clear acrylic; 1/4–20 four-arm knobs (8); 1/4" SAE flat washers (16); 1/4–20 flathead bolts, 2" long (12); 1/4–20 flathead bolts, 1 3/4" long (4); T-slot nuts (8); #6 x 1/2" panhead screws (4); #6 x 1" flathead wood screws (5); #8 x 1" flathead wood screws (26); #8 x 1 1/4" brass flathead wood screws (4); 3/4" T-track, 28" long (1); 3/4"-diameter 5 1/4"-long plastic knob extensions (2); 1/4–20 threaded rod 7" long (2); dust port (1).

board over the bit and snug the knobs. Slide a piece of the stock to be routed under the feather board, and press the feather board down on the stock so the feathers flex but the stock moves easily.

Tighten the knobs. Hold the stock against the fence, and feed it past the bit.

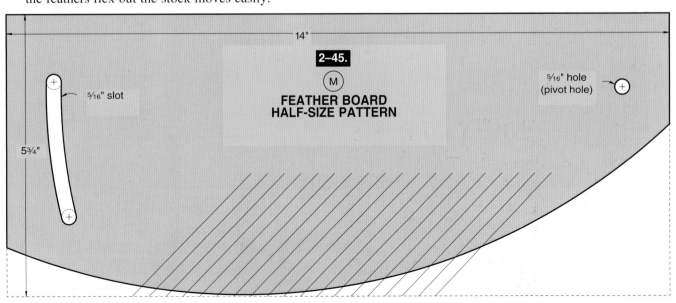

14"

5 3/4"

5/16" slot

5/16" hole (pivot hole)

2–45.

M

FEATHER BOARD
HALF-SIZE PATTERN

2–46.

This jig is a specialist at through or stopped dadoes and grooves.

DADO-CUTTING JIG

There are only a few things that can go wrong when you're routing dadoes for shelves. Unfortunately, they can ruin the job, so one of our freelance craftsmen, Erv Roberts, designed a router jig that's easy to align and cuts the exact thickness of the shelf (**2–46** and **2–47**). Best of all, the router can wander away from the fence while you're working—without ruining the job.

The secret is a flush-trimming pattern bit with a bearing above the cutters. As you guide the bit along a fence, the cut is directly below the bearing—one edge of the cut automatically falls

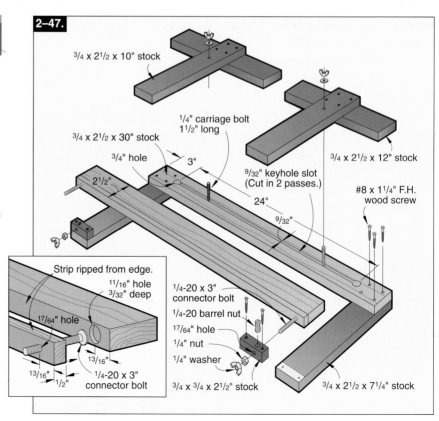

2–47.

3/4 x 2¹/2 x 10" stock

1/4" carriage bolt 1¹/2" long

3/4 x 2¹/2 x 30" stock

3/4 x 2¹/2 x 12" stock

3/4" hole

3"

9/32" keyhole slot (Cut in 2 passes.)

2¹/2"

24"

9/32"

#8 x 1¹/4" F.H. wood screw

Strip ripped from edge.

11/16" hole
3/32" deep

1/4-20 x 3" connector bolt

17/64" hole

1/4-20 barrel nut

17/64" hole

1/4" nut

13/16"

1/4" washer

13/16"

1/2"

1/4-20 x 3" connector bolt

3/4 x 3/4 x 2¹/2" stock

3/4 x 2¹/2 x 7¹/4" stock

exactly along the front of the fence. Erv's jig uses a router bit narrower than the dado you need, but has two fences. If your router wanders away from one, the other keeps you from routing outside your layout lines.

Begin by laying out both sides of the dado. Clamp the jig's fixed fence over one of the layout lines, and then slip the shelf stock between the fixed and movable fences. Turn the wing nuts to snug the movable fence against the stock, and clamp it in place. To make the cut, guide the bearing along one fence, and then the other, traveling clockwise. To make a stopped dado, use the optional stops.

Most of the jig screws together. Make the movable fence from two pieces, as shown in **2–47**. Slip the bolts through the holes in the narrow piece, and then glue the two pieces together. To form the shouldered keyhole slot in the fixed fence, drill the ¾" end holes, and then rout the ⁹⁄₃₂" slot with a ¼" bit chucked in your table-mounted router. Now, centering a ¾" bit on the end holes, rout a ³⁄₁₆"-deep shoulder in the fence's bottom face. The end holes allow you to add and remove the stops without taking off the wing nuts.

Do you need to rout a dovetail or keyhole slot? Simply install a guide bushing in your router subbase, and adjust the movable fence to the bushing's diameter.

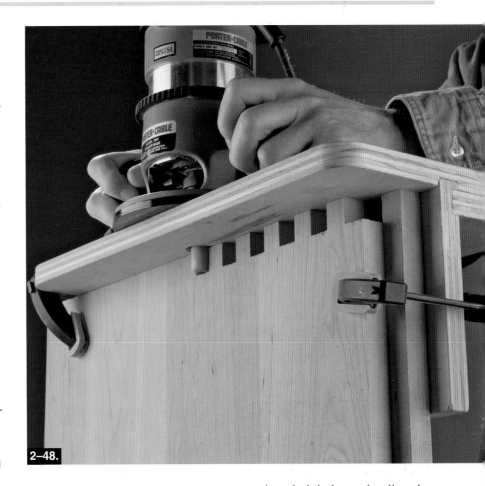

2–48.

BOX-JOINT JIG

Most box joints are cut on a tablesaw (see *page 90* for a tablesaw jig). But large projects, such as blanket chests, require a different approach. For really big boxes, we developed a clamp-on jig that works with a handheld router (**2–48** to **2–50**). It's simple to build and still easier to use.

First, Make the Large Parts

1 From ¾" plywood cut parts A, B, and C according to the Materials List. On the plate (A), mark the position of the ¼" carriage-bolt holes and radiused front corners, as shown on **2–50**. Also mark the location of the 1 x 3⅜" guide-bushing slot shown on **2–49**.

2 On the bottom surface of the plate, mark the location of the ¾"-wide, centered groove that runs the length of the plate. Then, mark another ¾"-wide groove on the vertical fence (B) located ⁷⁄₃₂" from its top edge, as shown on **2–50**.

3 Fit your router with a 1" straight bit. As shown on **2–49**, clamp a straightedge parallel to the marked slot and distanced from the center of the slot by half

2–49.

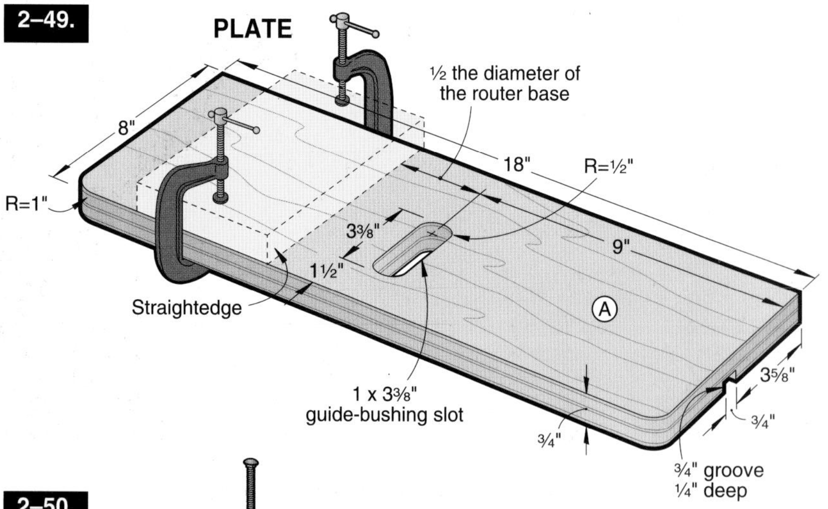
PLATE
½ the diameter of the router base
8"
R=1"
18"
R=½"
3⅜"
9"
1½"
Straightedge
Ⓐ
1 x 3⅜" guide-bushing slot
3⅝"
¾"
¾"
¾" groove
¼" deep

2–50.

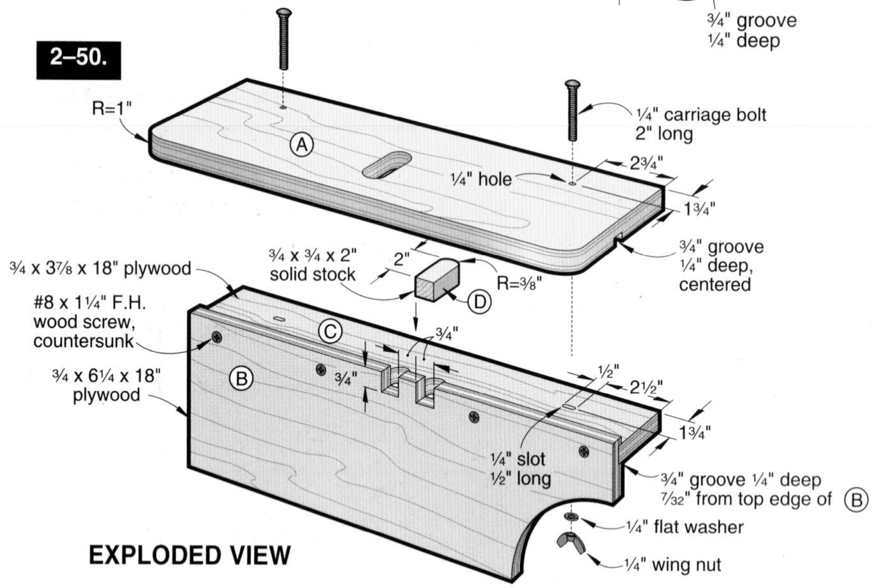
R=1"
Ⓐ
¼" carriage bolt
2" long
2¾"
¼" hole
1¾"
¾" groove
¼" deep,
centered
¾ x 3⅞ x 18" plywood
¾ x ¾ x 2"
solid stock
2"
R=⅜"
Ⓓ
#8 x 1¼" F.H.
wood screw,
countersunk
Ⓒ
¾"
¾ x 6¼ x 18"
plywood
Ⓑ
¾"
½"
2½"
1¾"
¼" slot
½" long
¾" groove ¼" deep
⁷⁄₃₂" from top edge of Ⓑ
¼" flat washer
¼" wing nut
EXPLODED VIEW

of the diameter of the router base. Rout the slot completely through the plate in several ⅛" passes for best results. The sides of the slot must be smooth, straight, and parallel.

4 Set up a dado blade to cut a groove that's ¼" deep and matches the thickness of your plywood. Cut the marked grooves on the plate (A) and vertical fence (B). You may need to slightly increase the width of the groove in part A so the edge of part B slides smoothly, but not loosely, in it.

Assemble the Jig

1 Glue and clamp the horizontal fence (C) into the groove in part B. Be sure the parts form a 90° angle. After the glue dries, drill and countersink ³⁄₃₂" pilot holes and secure the parts with #8 x 1¼" flathead wood screws.

2 Clamp the B/C assembly in your bench vise and place the plate (A) on top of it with the top edge of part B sitting in the groove in the plate. Align the part ends and clamp them together. Drill the marked ¼" carriage-bolt holes. Remove the plate, and elongate the two holes in part C ⅛" in both directions along the length of the jig.

3 Place the plate (A) atop the B/C assembly again, flush their ends, and tap the carriage bolts into their holes. Add the flat washers, and tighten with wing nuts.

4 Mount a 1"-O.D. guide bushing to the baseplate of your router. Add a ¾" straight bit (we recommend using one with a ½" shank), and adjust it so it sticks ¼" out of the guide bushing. Position the guide bushing in the slot in the plate (A), and rout through assembly B/C. Make progressively deeper cuts until the cut in the assembly is ¾" deep.

5 Remove the carriage bolts and plate (A). Mark on C the position of the slot that holds the guide pin (D) exactly ¾" from the slot you cut in the previous step (see **2–50**). Align the slot in the

Materials List for Box-Joint Jig

PART	FINISHED SIZE T	W	L	MATL.	QTY.
A plate	¾"	8"	18"	P	1
B vert. fence	¾"	6¼"	18"	P	1
C horiz. fence	¾"	3⅞"	18"	p	1
D guide pin	¾"	¾"	2"	M	1
E key attachment	¾"	1½"	2½"	P	1

Materials Key: P = plywood; M = maple or other dense wood.
Supplies: #8 x 1 ¼" flathead wood screws (6); ¼ x 2" carriage bolts (2); ¼" flat washers (2); ¼" wing nuts (2).

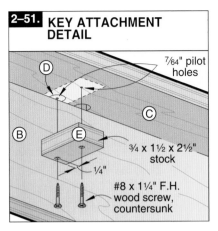

2–51. KEY ATTACHMENT DETAIL

- D
- C
- B
- E
- 7/64" pilot holes
- 3/4 x 1½ x 2½" stock
- ¼"
- #8 x 1¼" F.H. wood screw, countersunk

plate (A) atop assembly B/C so your router will cut a ¾"-deep slot on your marks. Clamp the plate to the assembly, and rout the slot just as you did in the previous step.

2–52.

Add the Guide Pin

1 From solid stock (we used maple), cut a ¾ × ¾ × 2" guide pin (D). Check its side-to-side fit in the second slot you cut in assembly B/C. It needs to be snug. Now, mark and cut its radiused end to fit in the slot. Place the pin in its slot. Do not glue it.

2 Cut part E according to the Materials List. Flip the jig over and position E over the radiused end of D. (See the Key Attachment detail in **2–51**.) Drill and countersink 7/64" pilot holes in E that go ½" deep into C and D. Attach E to C and D with #8 × 1¼" wood screws.

3 Sand all surfaces and sharp edges. That's it. Now, follow the text below to find out how to put this joint maker to work. Once you have this jig set up, you'll find it a breeze to make large boxes of many sizes and shapes.

Cut Large Box Joints

This jig cuts box joints with ¾ × ¾" pins and matching notches in stock that's ideally about ¾" thick (**2–52**). Although you can use the jig with stock of other thicknesses, we think ¾" pins and notches look best in ¾" stock.

Your panels can be any length and width; just remember to make them about 1" wider than their finished width. Doing that will allow you to trim them for evenly sized pins and notches at the top and bottom of each joint later.

Prepare Your Project Panels

Mark the top edges and face sides (the surfaces that will face outward in the assembled box) on each of the four panels. Number each of the adjoining panel ends so you can match up your adjoining pieces at any point in this machining process.

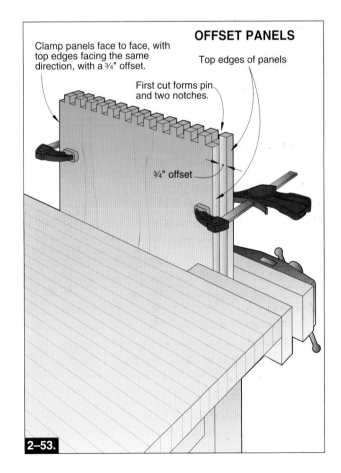

OFFSET PANELS

Clamp panels face to face, with top edges facing the same direction, with a ¾" offset.

Top edges of panels

First cut forms pin and two notches.

¾" offset

2–53.

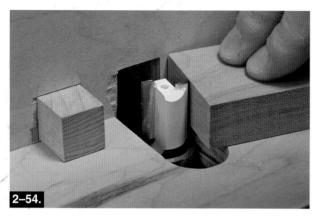

2–54.

Match the height of the router bit to the thickness of a scrap piece from your panel stock, and then raise the bit ¹⁄₃₂".

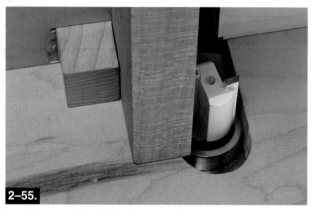

2–55.

The bit-to-guide-pin distance should exactly match a ¾"-thick piece of scrap.

Grab two panels with adjoining ends. Position the panels face side-to-face side, top edges facing the same direction. Align their ends flush and offset one panel ¾". (See **2–53**, which shows two panels after being cut for box joints.) Clamp the panels together and stand them upright so the ends are about chest high. (For large panels, you can stand the longer of the two clamped panels on the floor and hold them upright in a bench vise.)

Ready Your Router

Put a 1"-O.D. guide bushing in your router's base, and install a ¾" straight bit. Turn the router upside-down and set the jig on the router's base. Now, adjust the height of the bit using a scrap block that matches the thickness of your panel stock, as shown in **2–54**. Then, raise the bit ¹⁄₃₂".

Now, set the distance between the bit and the jig's guide pin to exactly ¾". A scrap of ¾"-thick stock works well for doing this, as shown in **2–55**. To adjust this spacing, loosen the wing nuts and slide the two horizontal plates of the jig ever so slightly. Retighten the wing nuts.

Use the Jig

Place the jig on top of the clamped panels with its guide pin touching the edge of the workpiece that's offset to your left. Clamp both ends of the jig to the panels. Place the router bit and guide bushing into the end of the jig's slot closest to you. Turn on the router, hold it against the left side of the slot, and push it forward to the other end of the slot to cut a pin and matching notches. Turn off the router and wait for the bit to stop spinning before lifting the router off the jig.

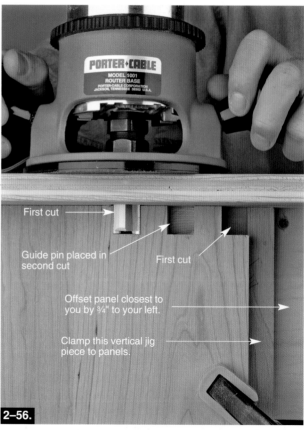

2–56.

Work from your left to right when cutting the pins and matching holes across the ends of two clamped panels.

Remove the debris from the jig and panel notches. Place the guide pin into the notch you just cut, and repeat the process, as shown in **2–56**, until you cut all the way to the other edge of the panel. For all of the cuts, remember to hold the bushing against the left side of the jig slot to help ensure consistent holes.

Check Your Results

You should be able to tap the box joints together by hand. If you have to force them together with a mallet, you may break a pin and the joints will be too tight to hold enough glue. Sloppy joints will prove weak and unattractive.

To fine-tune the box-joint fit, loosen the wing nuts and slide the guide pin closer to the bit in the case of too-tight joints. Sliding the guide pin further away from the bit will result in tighter joints. Be patient with this step; it may take several trials with very slight adjustments to get things right.

Also check that the ends of the pins stick slightly out of the notches so you can sand the joints flush. If the pins are short, increase the depth of the router-bit cut. In the event the pins are too long, decrease the cutting depth.

Trim the panels to width, being careful to leave evenly sized pins and matching notches at both panel edges. Add some glue and clamp two panels at a 90° angle, and then glue and clamp the other two panels. After the glue dries, glue and clamp these two assemblies.

2–57.

ROUTER-TABLE MULTI-JOINT JIG

Imagine machining stub and mitered tenons, mortises, and small box joints easily and accurately every time. With this hardworking jig and your table-mounted router (**2–57** and **2–58**), now you can. The fence guides the whole operation while the plunge and horizontal sliding fixtures move the workpiece up and down and across the router bit.

Beginning on *page 48*, see the numerous ways to put this handy jig to work in your shop. For even greater ease and accuracy in positioning the workpiece, consider adding an Incra Jig (available from woodworking supply firms) to the back side of the fence.

Make the Fence

1 Cut the fence upright (A) to the size listed in the Materials List from ¾" plywood.

2 Glue three pieces of ¾ × 2¼ × 29" plywood face-to-face with the edges and ends flush. Later, trim the block to 2" wide by 28" long to form the fence support (B).

3 To form the guides (C, I), cut three pieces of solid stock to ½ × 1 × 28". Cut or rout a ¼" rabbet ¼" deep along one edge of each strip. You'll use two of the guides (C) for the fence now; set the third piece aside for use later.

4 Keeping the top and bottom edges of the upright (A) flush with the outside edges of the guides (C), where shown on **2–59**,

clamp the guides to the upright.
Now, drill countersunk mounting
holes through each guide (C) and
into the upright (A), and screw
the guides to the upright.

5 With the bottom edges flush,
glue and clamp the support (B)
to the back side of the upright (A).

6 If you plan on attaching the
router jig to an Incra Jig
(which we recommend), drill a
pair of countersunk mounting
holes through the upright (A)
and support (B),
where shown
on **2–59**.

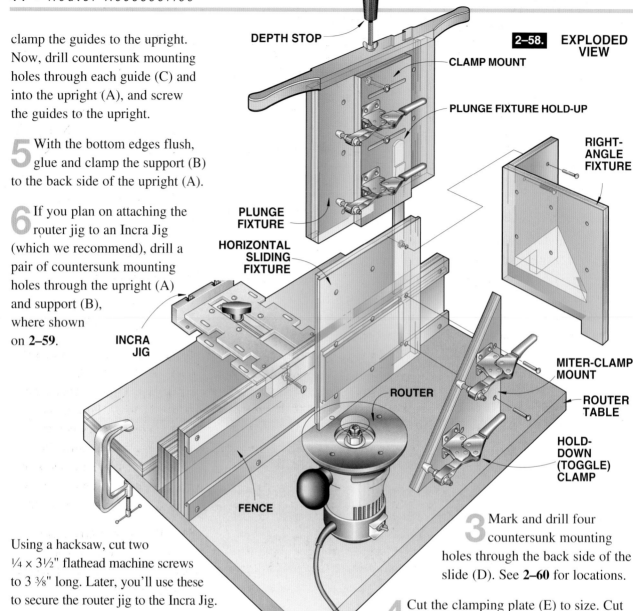

DEPTH STOP

2–58. EXPLODED VIEW

CLAMP MOUNT

PLUNGE FIXTURE HOLD-UP

RIGHT-ANGLE FIXTURE

PLUNGE FIXTURE

HORIZONTAL SLIDING FIXTURE

INCRA JIG

MITER-CLAMP MOUNT

ROUTER TABLE

ROUTER

HOLD-DOWN (TOGGLE) CLAMP

FENCE

Using a hacksaw, cut two
¼ x 3½" flathead machine screws
to 3 ⅜" long. Later, you'll use these
to secure the router jig to the Incra Jig.

Add the Horizontal Sliding Fixture

1 Cut the slide (D) to size from ½" solid stock.
The width of the slide should be just a hair under
the distance between the shoulders of the rabbets cut
in the guides (C) attached to the fence upright (A).

2 Cut or rout a ¼" rabbet ¼" deep along the edges of
the slide (D). Check the fit of the slide between
the guides (C). The slide must move back and forth
easily without slop for precision machining later. If
there is too much play, cut and rabbet another slide.

3 Mark and drill four
countersunk mounting
holes through the back side of the
slide (D). See **2–60** for locations.

4 Cut the clamping plate (E) to size. Cut
a ¼"rabbet along opposing edges. Each
rabbet must leave a ¼" lip, as noted on **2–60**.
Because some ¾" plywood is not exactly
¾" thick, machine the first cut a bit shy of a
½" depth and measure the remaining lip. Increase the
depth as necessary until a ¼" lip is left.

5 Mark the centerpoints for the nine ¹⁹⁄₆₄" holes on
the back side of the clamping plate (E), where
shown on **2–64** on *page 48*. For securing the clamp
mounts (G, L, N) to the clamping plates (E, H, K),
keep the holes exactly 3½" apart. To house the base
of the ¼–20 T-nuts, use a Forstner bit to drill a ¾"
counterbore ⅛" deep centered over each marked

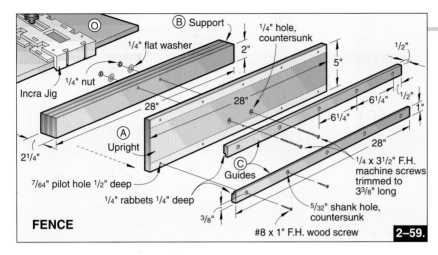

FENCE

Labels within figure:
- O — Incra Jig
- B Support
- 1/4" flat washer
- 1/4" nut
- 1/4" hole, countersunk
- 2"
- 5"
- 1/2"
- 1/2"
- 1"
- 28"
- 28"
- 6 1/4"
- 6 1/4"
- A Upright
- 2 1/4"
- 7/64" pilot hole 1/2" deep
- 1/4" rabbets 1/4" deep
- C Guides
- 3/8"
- 28"
- 1/4 x 3 1/2" F.H. machine screws trimmed to 3 3/8" long
- 5/32" shank hole, countersunk
- #8 x 1" F.H. wood screw
- **2–59.**

centerpoint on the back side of the plate. Drill a 19/64" hole centered inside each counterbore. (The Forstner bit leaves a slight depression that you can center your bit over.) Use a backing board to minimize chip-out.

6 Tap a 1/4–20 T-nut into each 3/4" counterbore on the back side of the clamping plate (E).

7 Clamp the slide (D) to the clamping plate (E), where dimensioned on **2–64**. Using the previously drilled holes in the slide as guides, drill pilot holes and drive the screws that secure the slide to the clamping plate pieces together. Verify that the sliding assembly (D/E) slides freely when slid in place between the fence guides (C). If the fit is too tight, use pieces of paper between parts D and E to act as spacers. Again, you'll want a smooth sliding action with slop.

8 To make the hold-up (F), enlarge and transfer the half-sized Front and Side View patterns from *page 47* to a piece of solid stock measuring 3/8 x 1 x 10". Drill the mounting holes, and then cut the hold-up to shape.

9 Screw the hold-up (F) to the clamping plate (E), where shown on **2–60** and **2-64**.

Make the Plunge Fixture

1 Cut the clamp mount (G) and clamping plate (H) to size.

2 Mark the three 5/16" slots and the locations for the two hold-down clamps on the clamp mount (G), where located on **2–65**, on *page 48*. Position the hold-down clamps against the marked locations, and transfer the mounting hole locations from the hold-downs to the clamp mount.

3 Drill blade start holes, and scroll-saw the three slots to shape. Drill the mounting holes for the hold-downs.

4 Mark and drill the T-nut holes in the clamping plate (H), using the same process you used to drill the holes in the horizontal-sliding-fixture clamping plate (E). Tap the T-nuts in place.

5 Crosscut the plunge-fixture guides (I) from the third piece of guide stock you cut earlier for the guides (C). Clamp (no glue) the guides to the back side of the clamping plate (H). Fit the plate/guide assembly (H/I)

2–60.

HORIZONTAL SLIDING FIXTURE

Labels within figure:
- F Hold-up
- 1 5/8"
- 1 5/8" rabbet 3/16" deep
- 3/8"
- 9/64" shank hole in F, mating pilot hole in part E is a 3/32" hole 1/2" deep
- 3/16" flat washer
- #6 x 3/4" R.H. wood screw
- #8 x 1" F.H. wood screw
- 1"
- 1/4"
- 1/4" T-nut
- 8 3/8"
- 8"
- 3 1/2"
- 3/16"
- 5/32" shank hole, countersunk on back
- 3/4"
- 1 1/2"
- D Slide
- 1/4" rabbets 1/4" deep
- 19/64" hole with a 3/4" counterbore 1/8" deep on back
- E Clamping plate
- 1/4" rabbet 1/2" deep
- 10"
- 10"

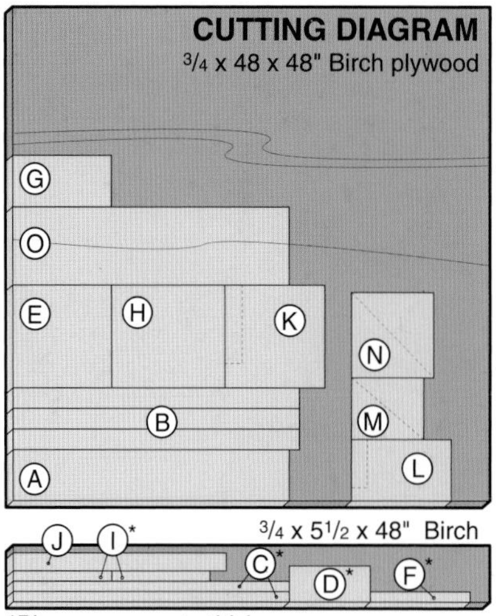

onto the horizontal sliding fixture. Adjust the location of the guides (I) if necessary for a good fit onto the rabbets on the ends of clamping plate (E). Drill mounting holes, and screw the guides (I) to the clamping plate (H).

6 Enlarge and transfer the handle pattern (J) in **2-61** to ¾" solid stock. Cut the handle to shape. Drill the mounting holes where marked. Rout a ³⁄₁₆" round-over along the top and bottom edges of the handle where marked on the pattern. Remove the paper pattern. Screw the handle to the top edge of the clamp plate (H).

7 Slide the plunge fixture (H/I/J) onto the horizontal sliding fixture (D/E/F). Verify that the hold-up (F) slides by the handle when the hold-up is held back slightly. The hold-up should support the plunge fixture above the tabletop.

Add the Right-Angle Fixture and the Miter-Clamp Mount

1 Using the Right-Angle Fixture drawing in **2–63** and **2-66** (*page 49*) insert for reference, cut the clamping plate (K), clamp-plate mount (L), and brace (M) to shape.

2 Drill the countersunk mounting holes and T-nut counter-bores, and screw the assembly together.

CUTTING DIAGRAM
¾ x 48 x 48" Birch plywood

¾ x 5½ x 48" Birch

*Plane or resaw to thickness listed in Bill of Materials.

Materials List for Router-Table Multi-Joint Jig

PART	FINISHED SIZE			MATL.	QTY.
	T	W	L		
FENCE					
A upright	¾"	5"	28"	BP	1
B* support	2¼"	2	28"	LP	1
C guides	½"	1"	28"	B	2
HORIZONTAL SLIDING FIXTURE					
D slide	½"	3½"	8"	B	1
E clamping plate	¾"	10"	10"	BP	1
F hold-up	⅜"	1"	10"	B	1
PLUNGE FIXTURE					
G clamp mount	¾"	5"	10"	BP	1
H clamping plate	¾"	10"	11½"	BP	1
I* guides	½"	1"	10"	B	2
J handle	¾"	1¾"	21⅝"	B	1
RIGHT-ANGLE FIXTURE					
K clamping plate	¾"	10"	10"	BP	1
L clamping-plate mount	¾"	6"	10"	BP	1
M brace	¾"	6"	7¼"	BP	1
MITER CLAMP MOUNT					
N mount	¾"	8¼"	8¼"	BP	1
INCRA JIG MOUNT					
O Mount	¾"	8"	28"	BP	1

*Cut parts oversized. Trim each to finished size according to the instructions.

Materials Key: BP = Baltic-birch plywood; B = birch; LP = laminated plywood.
Supplies: Twenty #8 x 1" flathead wood screws; eight #8 x ¼" flathead wood screws; two #6 x ¾" round-head wood screws with flat washers; eight #12 x ¾" panhead sheet metal screws; four ¼ x 1 ¾" panhead machine screws with flat washers; ¼" wing nut, ¼" nut, and ¼" flat washer; twenty-five ¼–20 T-nuts; two ¼ x 3 ½" flathead machine screws.
Buying Guide
Hardware kit. All the pieces listed in the supplies above plus two 2"-reach hold-down (toggle) clamps, and one 1¼" plastic knob with a mating piece of threaded rod 4" long, plus screws, washers and nuts listed above.
Supplies. Wood® Kit RJ1, $45 plus $4.75 shipping, ppd. Add $30 for two additional clamps and $39.95 for an Incra Jig. Schlabaugh and Sons Woodworking, 720 14th Street, Kalona, IA 52247 or call 800-346-9663.
Ready-to-assemble kit. All the pieces listed in the hardware kit above, plus all the Baltic-birch plywood pieces listed in the Materials List cut to size and shape with predrilled holes. Wood Kit RJ2, plus shipping (call for shipping charges). Schlabaugh and Sons Woodworking, address and phone above.

2–61. HANDLE HALF-SIZE PATTERN

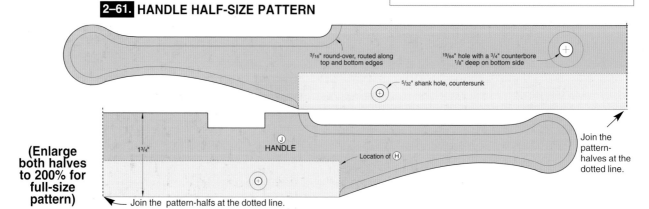

³⁄₁₆" round-over, routed along top and bottom edges

¹⁹⁄₆₄" hole with a ¾" counterbore ⅛" deep on bottom side

⁵⁄₃₂" shank hole, countersunk

1¾"

J
HANDLE

Location of H

Join the pattern-halves at the dotted line.

(Enlarge both halves to 200% for full-size pattern)

← Join the pattern-halfs at the dotted line.

3 Cut the miter-clamp mount (N) to shape. Mark the centerpoints for the three ¼" holes used for securing the mount to parts H and E later. Drill the holes.

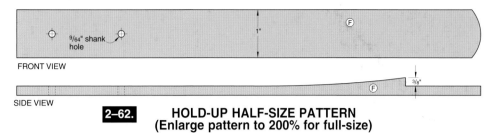

2–62. **HOLD-UP HALF-SIZE PATTERN**
(Enlarge pattern to 200% for full-size)

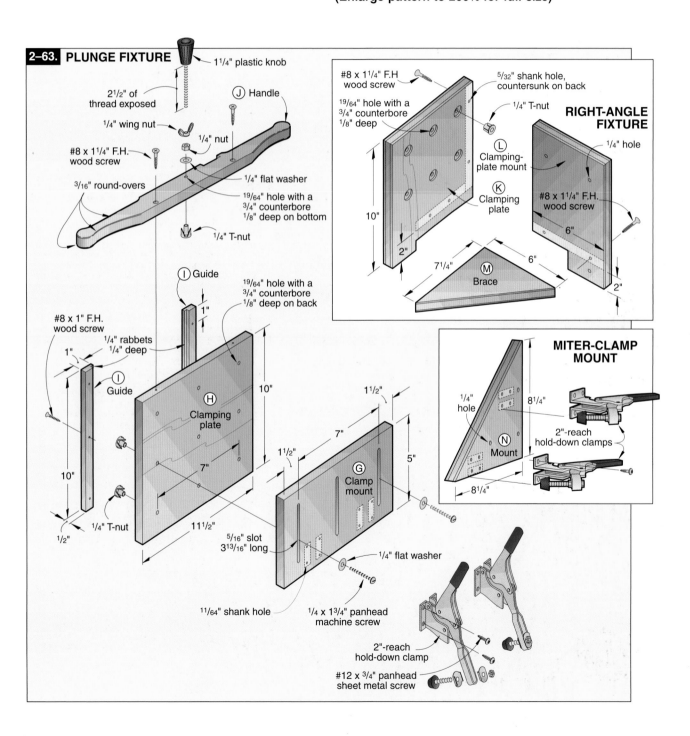

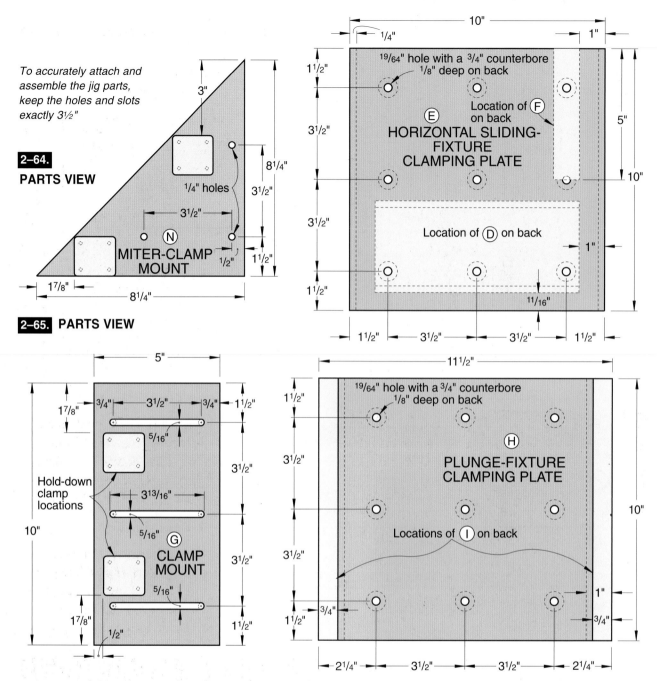

To accurately attach and assemble the jig parts, keep the holes and slots exactly 3½"

2–64. **PARTS VIEW**

3"

8¼"

3½"

¼" holes

3½"

N

MITER-CLAMP MOUNT

1⅞"

8¼"

½"

1½"

2–65. **PARTS VIEW**

10"

¼"

1"

1½"

¹⁹⁄₆₄" hole with a ¾" counterbore ⅛" deep on back

Location of **F** on back

E

HORIZONTAL SLIDING-FIXTURE CLAMPING PLATE

3½"

3½"

Location of **D** on back

5"

10"

1"

1½"

1½"

3½"

3½"

1½"

¹¹⁄₁₆"

5"

1⅞"

¾"

3½"

¾"

1½"

5⁄₁₆"

3½"

3 ¹³⁄₁₆"

5⁄₁₆"

G

CLAMP MOUNT

5⁄₁₆"

1⅞"

½"

3½"

1½"

Hold-down clamp locations

10"

11½"

1½"

¹⁹⁄₆₄" hole with a ¾" counterbore ⅛" deep on back

3½"

H

PLUNGE-FIXTURE CLAMPING PLATE

3½"

Locations of **I** on back

1½"

¾"

1"

¾"

2¼"

3½"

3½"

2¼"

10"

4 Position the hold-down clamps on the mount (N), and mark the mounting hole centerpoints. Drill the holes, and screw the hold-down clamps in place.

5 If you'll be using the Incra Jig, cut the Incra Jig mount (O) from ¾" plywood. Cut it 8" wide by the same length as your router table. Secure the jig to the mount (O) and to the fence support (B).

6 Remove all of the hardware (except the T-nuts) from the assemblies. Sand each smooth and add the finish. (We sprayed on a couple coats of lacquer.)

Set Up the Jig for Three Different In-Line Operations

1. End-Grain Grooves (2–67 and 2–68)

Clamp the workpiece with its end flush with the table surface. Then, slide the fence over until the router bit is centered on the thickness of the workpiece. (Generally for mortises, use a router bit with a diameter that's one-third the thickness of the workpiece.)

Now, clamp both ends of the fence to the table. Adjust the height of the bit to get the correct depth

2–66.

PARTS VIEW

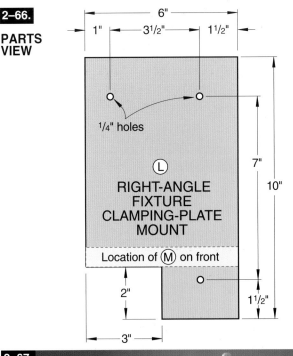

¼" holes

Ⓛ

RIGHT-ANGLE
FIXTURE
CLAMPING-PLATE
MOUNT

Location of Ⓜ on front

6"
1" 3½" 1½"
7"
10"
2" 1½"
3"

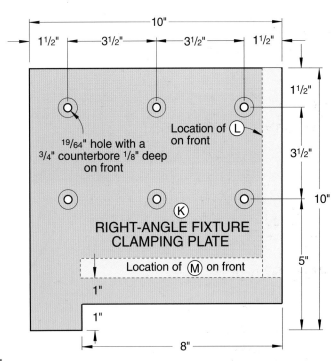

¹⁹/₆₄" hole with a
¾" counterbore ⅛" deep
on front

Location of Ⓛ
on front

Ⓚ

RIGHT-ANGLE FIXTURE
CLAMPING PLATE

Location of Ⓜ on front

10"
1½" 3½" 3½" 1½"
1½"
3½"
10"
5"
1"
1"
8"

2–67.

2–68.

for the groove. Raise the plunge fixture and set the depth stop so the bit will cut half the depth of the groove. Then turn the router on, and push the fixture and workpiece across the bit. Repeat with the workpiece positioned for a full-depth cut.

2. Stub tenons (2–69 and 2–70)

To make a stub tenon, follow the instructions for making an end-grain groove, only move the fence back so that the router bit cuts the outside face of the workpiece, as shown in **2–69**. Unclamp the workpiece, turn it around, and make another pass to

2–69.

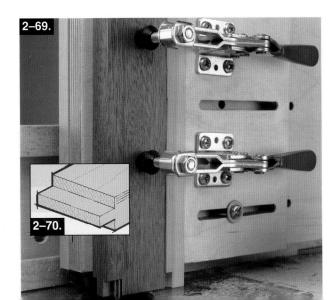

2–70.

2–71.

2–72.

45°

cut the other face. (Use multiple passes at different depths if necessary.) You also can cut a setback on the tenon by following the directions listed under the setup for right-angle operations on *page 51*.

3. Mitered Grooves and Tenons (2–71 and 2–72)

To make a 45° mitered version of the previous two cuts, simply remove the clamp mount and attach the miter-clamp mount to the plunge fixture. Then, follow the same procedures.

Three Plunge Cuts You Can Make

To get started in plunging operations, set up your jig the same way you did for in-line operations. You'll also need a pair of stops for the fence. We used small wooden handscrews.

1. Vertical Plunge Mortise

To cut a vertical plunge mortise (**2–74** and **2–75**), do the following:

2–73.

The tapered hold up enables you to secure the plunge fixture in the up position. To release, simply push out with your thumb.

2–74.

2–75.

1 Attach the clamp mount to the plunge fixture in the upright position, as shown in **2–74**, and clamp the workpiece to the plunge fixture. The bottom of the plunge fixture and the end of the workpiece should sit flush on the surface of the table. Adjust the height of the router bit to equal the depth of the mortise.

2 Next, raise the plunge fixture up so the end of the workpiece is just above the tip of the router bit. Adjust the fence so the bit is centered in the thickness of the workpiece. Clamp small handscrews or stops on the right and left sides of the fence to control the length of the mortise. Pull the plunge fixture up until the tapered hold-up clicks into place. Set the depth stop so that you cut one-third of the depth of the mortise.

3 Now, slide the plunge fixture against the right stop. Turn the router on, and disengage the hold-up by pushing it out with your right thumb. Lower the plunge fixture and workpiece onto the bit until the depth stop hits the top of the horizontal-sliding fixture. Then

slide the plunge fixture to the left until it butts against the stop.

4 Raise the plunge fixture until it engages the hold-up. Reset the depth stop so that your next plunge cut will equal two-thirds of the depth of the mortise. Now, repeat the plunging operation. On the third or final plunge, reset the depth stop so that the plunge fixture and the end of the workpiece come down flush with the surface of the router table.

2. Horizontal Plunge Mortise

To make this cut, follow the same procedures as you did for making a vertical plunge mortise. But attach the clamp mount and the workpiece in the horizontal position, as shown in **2–76**. To make the loose tenon that joins the two mortises, cut a piece of stock as wide as the mortise is long and as thick as the mortise is wide. Then, round-over the corners with a router and a round-over bit to match the profile of the mortise. You can make these loose tenons in long pieces and then crosscut them to the size you need for the individual joints.

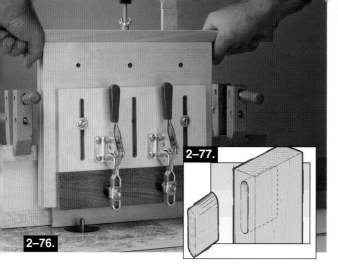

2–77.

2–76.

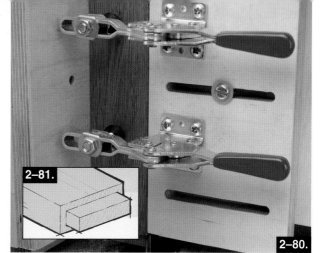

2–81.

2–80.

3. Miter Plunge Mortise

Again, follow the same set-up procedures for making a vertical plunge mortise, but only this time attach the miter-clamp mount (**2–78** and **2–79**). Be sure not to rout the mortise so deep that the bit penetrates the edge of the workpiece near the outside edge of the miter joint.

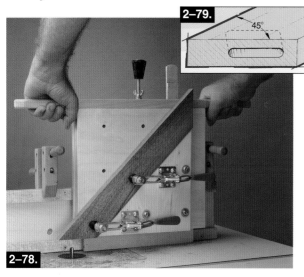

2–79.

45°

2–78.

Make These Two Joints with the Right-Angle Fixture

For right-angle operations (those operations where the face of the workpiece travels perpendicular to the router-table fence), you need to mount the right-angle fixture to the horizontal sliding fixture. Use a pair of ¼ x 1¾" panhead machine screws to mount the right-angle fixture to the horizontal fixture. Then, attach the clamp mount to the front face of the right-angle fixture, again using two ¼ x 1¾" panhead machine screws.

1. Tenon Setbacks

Place the workpiece against the front face of the right-angle fixture, as shown in **2–80**, and slide the clamp mount over until the workpiece is trapped between the clamp mount and the face of the horizontal sliding fixture. Secure the workpiece with the toggle clamps.

Now, raise the router bit to the length of the tenon and position the fence so the bit will remove the correct amount of material from the edge of the tenon cheek. Turn the router on and push the fixtures and workpiece across the bit. Repeat on the other edge of the workpiece.

2. Box Joints

Teamed with an accurate positioning device, such as the Incra Jig, our right-angle fixture will cut box joints easily. Set up the right-angle fixture and workpiece the same as for routing tenon setbacks, as shown in **2–82**. Now, after routing the first recess, move the fence back a distance equal to twice the diameter of the router bit, and rout the next recess. Continue until you reach the opposite edge of the workpiece. Rout the mating workpiece in the same manner, but remember to offset the first recess so that you end up with interlocking pins and recesses.

2–82.

DADO-CUTTING STRAIGHTEDGE

If you've ever put off building a project because it called for extra-long or difficult-to-do dadoes, wait no longer. Instead, let our router straightedge (**2–83** and **2–84**) come to your rescue. With it, you easily can rout one or a dozen dadoes, stopped or not. Edge-joining with your router also becomes a snap.

Note: The finished width of the base (A) will depend on the size of the plastic subbase of your router and the size of the straight bit you use. We cut our base extra wide and then trimmed it to finished width later with a ¾" straight bit.

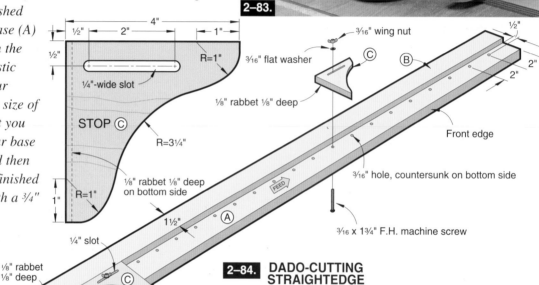

2–83.

2–84. **DADO-CUTTING STRAIGHTEDGE**

ROUTER TRAMMEL

For a few dollars and in just an hour or so, you can turn your router into a motorized compass (**2–85** and **2–86**). To adjust the size of circles it can cut, just move the pivot hole. Size the hole at the router end to accommodate the bit you'll be using.

2–85.

ROUTER TRAMMEL **2–86.**

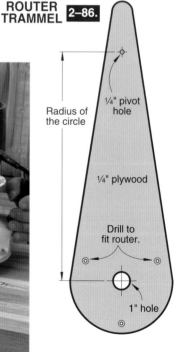

POCKET-HOLE ROUTING JIG

G iven how specialized the tools are for pocket-hole joinery, some woodworkers have balked at using the technique for projects. Nowadays, the dedicated jigs for drilling pocket holes are very affordable, but *WOOD*® magazine reader David Brunson of Loudon, Tennessee, came up with a router table jig for the same purpose. We tweaked his idea to get the design shown.

The jig (**2–87** and **2–88**) is simply a plywood platform supported on wedge-shaped runners cut at a 15° angle. One runner has a bottom cleat that rides in the miter-gauge slot of the router table and is controlled by a stopblock. This keeps the jig tracking in a straight line when routing, letting you make functional pockets. The hollow split-spring pin allows you to drill the centered pilot holes for screws.

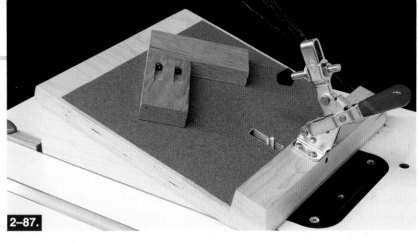

2–87.

On the jig's lower end, a wood fence acts as both a rest to locate the workpiece and a guide for drilling. Sandpaper and a toggle clamp grip the workpiece.

The pocket itself is cut with a ⅜" straight router bit. Simply raise the bit to 1¼" above the table surface, and slowly feed the jig, high end first, toward the cutter. It will plow into the underside of the platform and then emerge through the top. When the leading edge of the bit meets the index line marked ⅝" from the fence, shut off the router. Then clamp a stopblock in the miter gauge slot.

Before unclamping the workpiece, drill a ⁵⁄₃₂" hole through the metal guide (the ¼"-O.D. split-spring pin) in the fence, and into the workpiece.

2–88.

EXPLODED VIEW

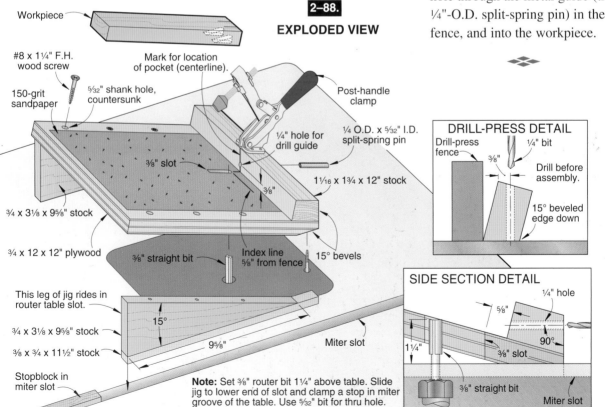

Workpiece

#8 x 1¼" F.H. wood screw

150-grit sandpaper

⁵⁄₃₂" shank hole, countersunk

Mark for location of pocket (centerline).

Post-handle clamp

¼" hole for drill guide

¼ O.D. x ⁵⁄₃₂" I.D. split-spring pin

1¹⁄₁₆ x 1¾ x 12" stock

⅜" slot

⅜"

¾ x 3⅛ x 9⅝" stock

¾ x 12 x 12" plywood

⅜" straight bit

Index line ⅝" from fence

15° bevels

This leg of jig rides in router table slot.

¾ x 3⅛ x 9⅝" stock

⅜ x ¾ x 11½" stock

Stopblock in miter slot

15°

9⅝"

Miter slot

DRILL-PRESS DETAIL

Drill-press fence

¼" bit

⅜"

Drill before assembly.

15° beveled edge down

SIDE SECTION DETAIL

¼" hole

⅝"

90°

1¼"

⅜" slot

⅜" straight bit

Miter slot

Note: Set ⅜" router bit 1¼" above table. Slide jig to lower end of slot and clamp a stop in miter groove of the table. Use ⁵⁄₃₂" bit for thru hole.

JIGS FROM THREE ROUTER EXPERTS

A trio of router experts share the shop-made jigs that they've found to be super handy over the years, along with some router wisdom.

Pat Warner's Two-Part Dado Jig

2–89.

Pat Warner.

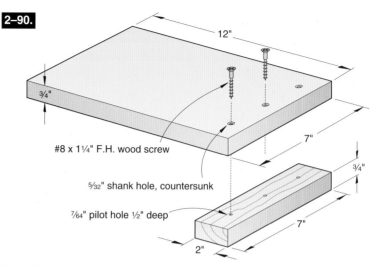

2–90.

12"

3/4"

7"

3/4"

#8 x 1¼" F.H. wood screw

⁵⁄₃₂" shank hole, countersunk

⁷⁄₆₄" pilot hole ½" deep

7"

2"

In addition to writing four router books, Pat Warner (**2–89**) has designed specialty router bits and is currently developing a collection of inexpensive, disposable, single-flute mortising bits.

Why You Need This Jig

Accuracy in routing requires attention to detail—and not much sophistication, according to Pat. This Californian's simple jig (**2–90**) helps match dado width to shelf thickness. Using Pat's two-part jig, you can set up to cut a dado in less time than it takes to equip a tablesaw with a dado set.

How to Build the Jig

You'll need to assemble two of the jigs, shown in **2–90**. (We cut two pairs of jigs, one from ¾"-thick plywood and another using medium-density fiberboard [MDF]). If you plan to make dadoes 10" or longer, build additional pairs with the top pieces at least 12" long or more.

Put the Two-Part Jig to Use

Before you begin cutting dadoes, you'll need a sample of the stock that the dado will ultimately hold in your project. Here's one key to a snug dado: Go through each sanding step you plan to follow until your sample piece reaches its finished thickness.

For this example, we're cutting a dado in the side of a cabinet for a shelf. Position the two-part jig where you plan to cut the dado; then snug the sanded shelf scrap

PERFECT DADOES: BEFORE AND AFTER

2–91.

After locating the dado position, place a piece of finish-sanded scrap between the jig's two parts. Then clamp the parts to the workpiece.

2–92. Dado

If the dado is too snug (sometimes caused by a pattern-cutting bit that's smaller than the bearing), shim out the shelf scrap with a sheet or two of paper, adjust the jig parts, and rout the dado again.

2–93. DADO JIG SIDE VIEW

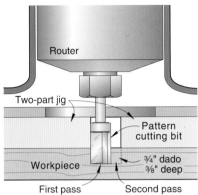

between the parts, as shown in **2–91**. Next, clamp both parts firmly in place at the edges away from the gap to keep the clamps from interfering with the router. After removing the sample (save this piece for future reference), set your router cutting depth equal to the thickness of the jig parts plus the depth of cut you want. Then rout a dado using a pattern-cutting bit (also sold as a shank-bearing guided trimmer) (**2–92**). The bearing of this bit rides along the edge of the jig parts and produces a crisp, square dado, as shown in **2–93**.

Be sure the cutter isn't larger than the bearing, or you'll tear up the edges of your jig parts.

Patrick Spielman's Multi-Hole Doweling Jig

Patrick Spielman (**2–95**), of Fish Creek, Wisconsin, has authored more than 75 woodworking books. His original *Router Handbook* sold more than a million copies, and his revised *New Router Handbook* remains one of this publisher's best-selling wood-working titles. Patrick shares with us his multi-hole doweling jig.

Why You Need This Jig

In addition to positioning dowels for most doweling joints (we found the jig particularly useful for face-frame joinery), this is a great jig for aligning shelf-support pin holes.

The dowel joint makes a lot of sense to Pat, a former schoolteacher, because dowels provide extra mechanical strength when joining end to edge grain, and they're quicker to make than hand-cut dovetails. The dowel joint is pretty much foolproof and, with this jig, you can take the joint further and have the dowels come through the other side. The 2" counterbored slots in the ⅜"-thick plastic jig make Patrick's jig versatile. The adjustable, removable stop will help you precisely position face stock.

How to Build Patrick's Jig

Follow **2–100** to build the jig. Lay out and drill the ⅝" holes as

2–95.

Patrick Spielman.

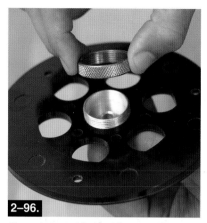

2–96.

Two-piece guide bushings easily attach to your router's base.

2–97.

Upcut spiral router bit.

ROUTING FACE-FRAME HOLES

2–98.

*Patrick Spielman's doweling jig makes quick work of doweling tasks, including the face frame joinery (end to edge grain) above. The ⅝" guide-bushing hole is compatible with ¼", ⁵⁄₁₆", or ⅜" dowel pins. The positioning block ensures quick, accurate alignment of the stock. **Illustrations 2–101** and **2–102** show routing dowel-pin holes in the rail and stile.*

POSITIONING SHELF-PIN HOLES

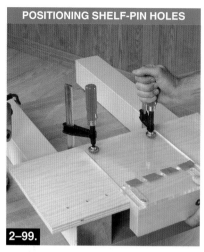

2–99.

Originally designed for doweling, this jig also helps you make evenly spaced holes for shelf pins. To avoid goofs, tape over holes you won't use, as shown above. (Spacing of 1¼–2" between holes works well in most applications.)

accurately as possible in the plastic. (We used a drill press for this step.)

Put the Jig to Use

With a guide bushing in your router, dowel holes always line up regardless of how accurately you spaced the ⅝" holes for your jig. For this type of plunge routing, use upcut spiral bits like the one shown in **2–97**. To position the jig and router to make face-frame holes and shelf-pin holes, see **2–98** and **2–99**.

To position the jig and router to make identically spaced dowel holes in face-frame stiles and rails, see **2–101** and **2–102**.

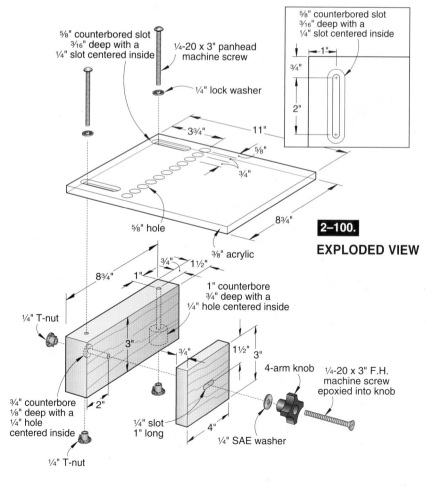

2–100.

EXPLODED VIEW

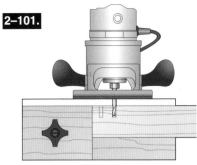

CUTTING HOLES ON STILE EDGES

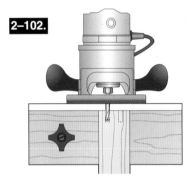

CUTTING HOLES ON RAIL ENDS

SHOP TIP

The Plastic World

With more than 35 years experience with routers, Patrick Spielman has designed several jigs incorporating plastic parts. We built Patrick's jig from ³/₈" polycarbonate because it's commonly available at hardware stores and home centers. However, he prefers ¹/₂" polycarbonate, a thickness that's tougher to find.

But if you can find ¹/₂" plastic, you'll have fewer problems with guide bushings being too long for the router bits. Plus, thicker plastic makes it easier to rout adjustable slots, according to Patrick.

Patrick avoids acrylic plastic because it's brittle, and he's had fewer problems with polycarbonate plastic cracking or breaking around a hole or near the edge. We found a local polycarbonate supplier under "plastics" in the phone directory. The local price was about $2.90 per square foot for ³/₈"-thick polycarbonate cut to order.

Carol Reed.

Carol Reed's Finger-Saving Devices

Carol Reed (**2–103**) sees herself as a teacher of beginners. She's taught woodworking and router techniques for 20 years. This Phoenix-area resident also demonstrates her talents at woodworking and home and garden shows. Her first book, *Router Joinery Workshop*, was published in 2003.

Why You Need These Jigs

Carol likes to call these pushsticks (**2–104** and **2–105**) her success devices. She believes that using these pushsticks will not only make you safer, you'll reduce burn marks and errors. Overall, you'll enjoy more success.

How to Build Carol's Jigs

Follow **2–104** and **2–105** to cut the pieces for Carol's two router table accessories. You'll find the handle pattern on *page 58*. Create a template of the handle design, and then make an armload of them. That way, you won't feel bad when you chew up one of your jigs.

We made our handles from easily worked and inexpensive MDF. To assemble the vertical

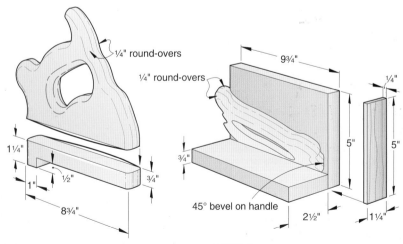

2–104. **NARROW STOCK PUSHSTICK**

2–105. **VERTICAL STOCK PUSHSTICK**

PUTTING PROBLEM-SOLVING PUSHSTICKS TO WORK

With the aid of a feather board, Carol Reed's pushstick for narrow stock keeps workpieces flat against the router table.

Carol's vertical pushstick, with a replaceable ¼" plywood heel, helps you press tall pieces firmly against your router table fence.

pushstick used for routing the ends of long stock, glue and clamp the base to the MDF upright piece. After applying glue to the handle,

rub it across the upright and clamp it in place. Glue on the heel last.

To make the pushstick for narrow stock, glue the base to the

handle and clamp until dry. For safety, do not use metal fasteners to hold the pieces together.

Put the Pushsticks to Use

Paired with a feather board, Carol's narrow stock pushstick stick, shown in **2–106**, supports the ends of tall pieces.

Carol uses her handle design at the tablesaw and jointer, too (**2-107**). After you chew up the sole and heel of the pushstick, send it through the jointer and attach new pieces.

And don't think of using just ¾"-thick-material to make the pushsticks. Carol has safely routed with ⅜" pushsticks and a little heel.

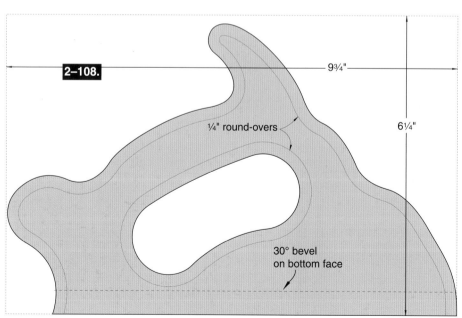

HALF-SIZE HANDLE PATTERN
(Enlarge to 200% for full-size pattern.)

CHAPTER 3

Treat Your Tablesaw

T HE TABLESAW IS THE REAL *workhorse of any woodworking shop. You undoubtedly use yours more than any other tool that you own, and most likely are aware of all that it can and can't do.*

In this chapter, you'll definitely find jigs, fixtures, and accessories that you can economically build to not only increase your saw's list of "can dos," but also make performing them easier, faster, and more accurate. So no matter the type of tablesaw you have, the following pages offer a wide selection of projects, both large and small, to help that workhorse do even more, and do it better. And you can begin with the easy jig that follows.

STRAIGHT-EDGE CUTTING JIG

Attempting to rip a straight edge along a board with irregular edges can be dangerous or downright impossible. One solution is to tack a straight board to the irregular board with finishing nails. But unfortunately, this method leaves small nail marks in the top surface of the workpiece.

So try this method: Construct a carrier board (**3–2**) from ¾" plywood to a width and length to accommodate most of your boards (14" × 7' works fine in most cases). As shown in **3–1**, you can quickly clamp the workpiece to this carrier board, and then rip one edge. Remove the workpiece from the carrier board, place the jig aside, and position the just-ripped edge along the fence to straighten the other edge.

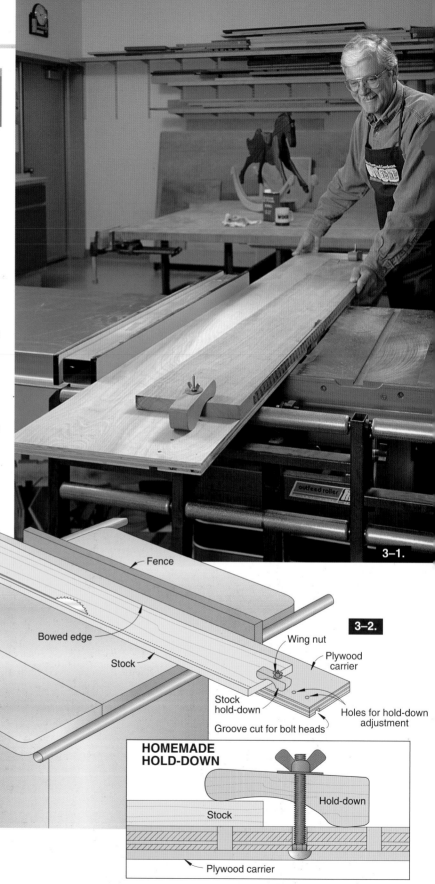

3–1.

Fence

Bowed edge

Stock

Wing nut

Plywood carrier

Stock hold-down

Holes for hold-down adjustment

Groove cut for bolt heads

3–2.

HOMEMADE HOLD-DOWN

Hold-down

Stock

Plywood carrier

ROCK-SOLID OUTFEED TABLE

I f you've had much experience using commercially available roller-stands as outfeed tables, you're already well aware of their shortcomings. Namely, they like to tip over. And, unless you precisely align the roller(s), they tend to track the workpiece off to one side or the other. Add to that the cost and limited uses for roller-stands, and we believe there has to be a better way!

Here's a no-cost solution that works in conjunction with what we consider to be one of the most versatile tools for any workshop: a Black & Decker Workmate. Simply join two pieces of leftover ¾" plywood to form a "T," as shown in **3–4**. If desired, you can apply paraffin wax or plastic laminate to the top surface, allowing your workpiece to slide more smoothly. Or, build the top surface from a piece of melamine-coated particle-board or kitchen countertop if you have a scrap handy.

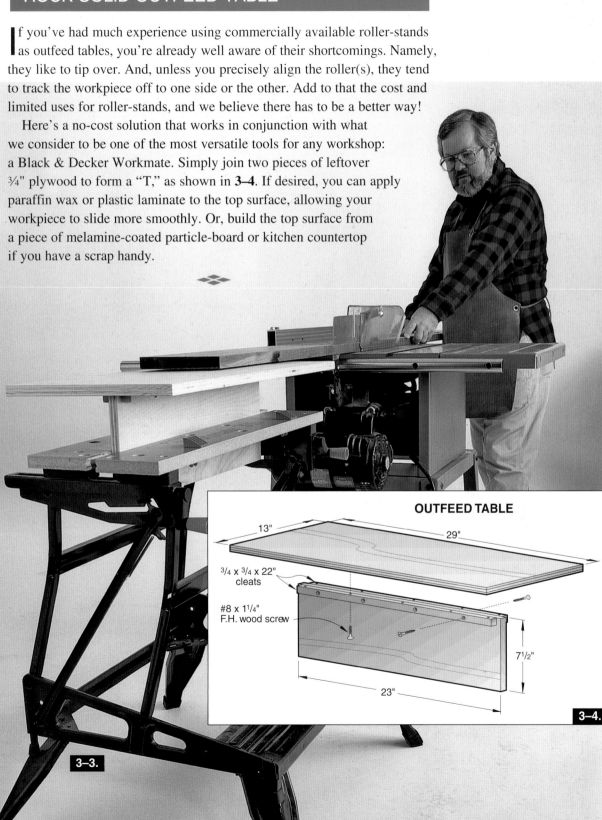

3–3.

OUTFEED TABLE

13"

29"

¾ x ¾ x 22" cleats

#8 x 1¼" F.H. wood screw

7½"

23"

3–4.

3–5a.

First, Size the Table to Fit Your Saw

If your tablesaw fence clamps to the rear rail, as shown in **3–6**, you'll need to leave a gap between the outfeed and saw tables for the fence to travel in. To calculate the gap width, measure from the back of your saw table to the rearmost part of the fence, and add ¼".

You'll also need to size the fixed table, which should be just wide enough for the extension table to clear the motor when folded down. On belt-driven saws, crank the blade all the way down to find the maximum reach of the motor, and then measure the horizontal distance between the back of the saw table and the back of the motor. If you have to

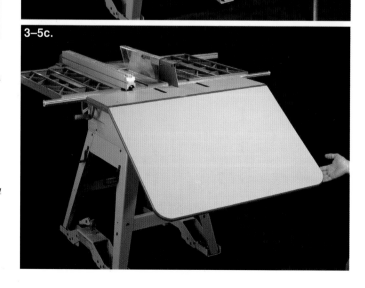

3–5b.

3–5c.

FOLDING OUTFEED TABLE

This handy folding fixture (**3–5 a-c**) adds only inches to your saw when stored, but provides more than 3' of stock support beyond the blade for safer and more convenient cutting.

(***Note:*** *This fixture attaches to any stationary tablesaw, regardless of the style or material of your saw's extension wings. We built ours to fit a Ridgid TS2424 tablesaw with webbed, cast-iron extension wings, but we'll also show you how to mount the outfeed table to solid cast-iron, stamped-steel, and shop-built wooden tables.*)

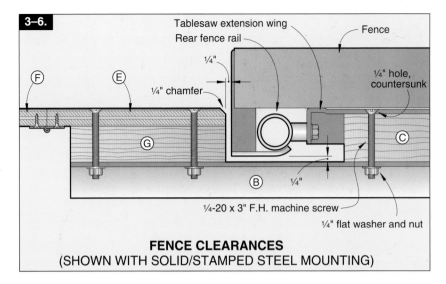

3–6.

Tablesaw extension wing

Rear fence rail

Fence

¼"

¼" hole, countersunk

¼" chamfer

F

E

C

G

B   ¼"

¼-20 x 3" F.H. machine screw

¼" flat washer and nut

FENCE CLEARANCES
(SHOWN WITH SOLID/STAMPED STEEL MOUNTING)

leave a gap for the fence, subtract the width of the gap to find the fixed table's length.

Finally, measure from the front of your saw's cabinet to the back edge of its table. Add the motor clearance measurement you just took, and then subtract 1" to find the length of the mounting bars.

Install the Mounting Bars

1 Size the mounting plates (A), if needed, to straddle three cells (as shown in the Web Mounting detail in **3–10**, on *page 65*).

2 Cut the mounting bars (B) to length. To locate the mounting bars, crank the saw blade to full height and adjust the bevel to 45°. Using a level, mark a minimum clearance line on the back edge of the saw table, plumb above the farthest reach of the motor.

3 Using **3–6** and the Web Mounting detail in **3–10** as a guide, cut two spacer blocks (C) and filler blocks (D), if needed, to size. For a rear-locking fence, the spacer blocks must be thick enough to allow ¼" clearance between the mounting bar and the fence mechanism.

4 Temporarily attach the spacer blocks to the filler blocks, steel bars, and mounting bars with cloth-backed double-faced tape. Drill and countersink mounting holes through the taped stack.

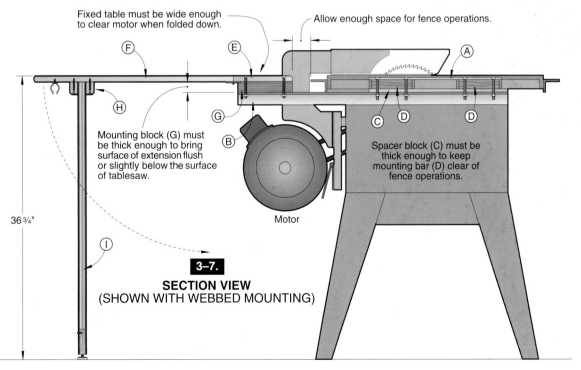

Fixed table must be wide enough to clear motor when folded down.

Allow enough space for fence operations.

F

E

A

H

G

B

C   D   D

Mounting block (G) must be thick enough to bring surface of extension flush or slightly below the surface of tablesaw.

Spacer block (C) must be thick enough to keep mounting bar (D) clear of fence operations.

36¾"

I

Motor

3–7.

SECTION VIEW
(SHOWN WITH WEBBED MOUNTING)

3–8.

Transfer miter gauge slot locations from bar to fixed table.

Ⓔ

MARKING THE MITER-GAUGE SLOT LOCATION

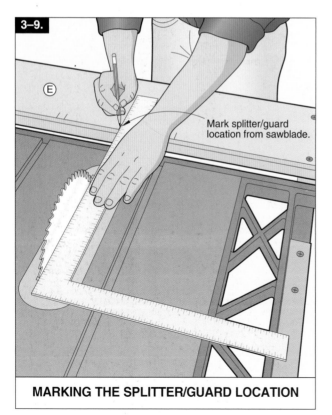

3–9.

Ⓔ

Mark splitter/guard location from sawblade.

MARKING THE SPLITTER/GUARD LOCATION

For non-webbed tables, clamp the spacer blocks and mounting bars in place on the saw and drill up from the bottom; then countersink the top. Remove the tape and bolt the assemblies in place.

Make and Mount the Fixed and Extension Tabletops

1 From ¾" medium-density fiberboard (MDF) or plywood, cut the fixed table (E) and extension table (F) to the sizes listed in the Materials List on *page 66*. Shape the radius on two corners of part F.

2 Cut top and bottom pieces of plastic laminate 1" longer and wider that each table segment. (We used a less-expensive type of laminate, called balance sheet, for the bottoms.)

3 Adhere the laminate to both sides of parts E and F with contact cement, and then trim the laminate using a flush-trim bit in your router. Chamfer and round over the edges, as shown in **3–6** and the Leg Mount detail in **3–10**, and paint, if desired.

4 Lay the fixed table on the mounting bars and place a straightedge on your saw top overhanging the fixed table. Measure between the table top and the straightedge and subtract ¹⁄₁₆". Cut the mounting blocks (G) to this thickness. Finally, cut the mounting blocks 1" shorter than the length of part E.

5 Clamp the fixed table and mounting blocks in position on the mounting bars. Drill and counter-sink screw holes, and bolt the assembly together.

6 Locate the miter-gauge slot extensions and splitter/guard slot, as shown in **3–8** and **3–9**. Remove the fixed table and dado the miter slots a little wider and deeper than your saw's slots so the bar clears easily. In each slot, drill a ¾" dust-escape hole, centered in the slot and 1½" from the back edge of the fixed table. Cut the splitter/guard slot, remembering to shape one side so the splitter will have room to move during bevel-cutting operations. Check the fit of the splitter/guard at 90° and 45°, and cut more clearance if needed.

3–10.

EXPLODED VIEW

7 Butt parts E and F together face down on your workbench. Install the continuous hinge.

8 Cut the leg mount block (H) to size and drill a 1" hole in the center. Attach the leg mount to the extension table and chamfer the edges, as shown in the Leg Mount Detail in **3–10**. Mount broom clips 2½" from the back and 6" from each side to store the leg when not in use. Bolt the outfeed table assembly to the mounting bars.

Make the Leg

1 With the extension table at full height, measure from the bottom of the extension to the floor. Subtract 1" for the adjustable

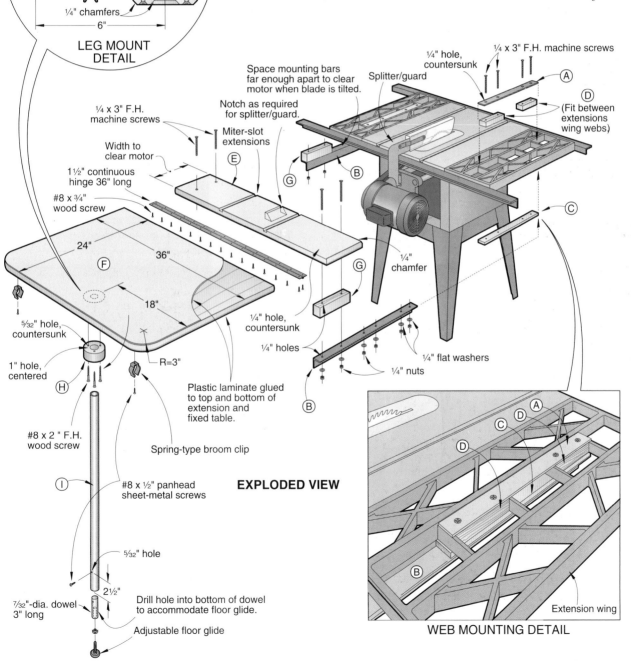

½" round-over bit centered on edge (Rout top and bottom edges.)

Spring-type broom holder

(H)

¼" chamfers

6"

LEG MOUNT DETAIL

¼ x 3" F.H. machine screws

¼" hole, countersunk

Splitter/guard

(A)

(D)

(Fit between extensions wing webs)

Space mounting bars far enough apart to clear motor when blade is tilted.

Notch as required for splitter/guard.

¼ x 3" F.H. machine screws

Miter-slot extensions

Width to clear motor

(E)

1½" continuous hinge 36" long

(G)

(B)

#8 x ¾" wood screw

24"

36"

(F)

18"

R=3"

¼" chamfer

(G)

⁵⁄₃₂" hole, countersunk

1" hole, centered

(H)

¼" hole, countersunk

¼" holes

¼" flat washers

¼" nuts

(B)

Plastic laminate glued to top and bottom of extension and fixed table.

(B)

#8 x 2 " F.H. wood screw

Spring-type broom clip

(I)

#8 x ½" panhead sheet-metal screws

EXPLODED VIEW

⁵⁄₃₂" hole

2½"

⁷⁄₃₂"-dia. dowel 3" long

Drill hole into bottom of dowel to accommodate floor glide.

Adjustable floor glide

(A)

(C) (D)

(D)

(C)

(B)

Extension wing

WEB MOUNTING DETAIL

Materials List for Folding Outfeed Table

PART	FINISHED SIZE			MATL.	QTY.
	T	W	L		
A** mounting plates	1/8"	1½"	*	S	2
B mounting bars	1½"	1½"	*	A	2
C* spacer blocks	*	1½"	*	SS	2
D** filler blocks	*	1½"	*	SS	4
E* fixed table	¾"	36"	*	C	1
F extension table	¾"	36"	24"	C	1
G mounting blocks	*	1½"	*	SS	2
H leg mount	1½"	3" diameter		SS	1
I leg	1" diameter	*		AT	1

*Size to fit your size, see text.
**For webbed extension wings only.

Materials Key: A = ⅛" angle iron; S = steel bar stock; SS = solid stock; C = choice of plywood or medium-density fiberboard (MDF); AT = aluminum tubing.

Supplies: Plastic laminate; contact cement; ⅛ x 1½" steel bar**; cloth-backed double-faced tape; 1½ x 36" continuous hinge; adjustable floor glide; spring-type broom clips (2); ⅞ x 3" dowel; ¼-20 x 3" flathead machine screws, flat washers, and nuts; #8 x ¾" flathead wood screws; #8 x ½" panhead sheet-metal screws.

floor glide, and cut a piece of 1" aluminum tubing (I) to that length.

2 Cut a 3" length of ⅞" dowel (you may need to sand down a 1" dowel) to fit inside the aluminum tubing, and drill a hole to accept the floor glide insert in the center of one end. Slip the dowel in one end of part H and drill a ⁵⁄₃₂" hole, where shown. Secure the dowel with a #8 panhead sheet-metal screw, and then press the insert and floor glide in place.

3 Now slip the leg into its mount on the extension table and double-check its length. Trim to fit or adjust the glide as necessary.

UNIVERSAL JIG WITH RIGHT-ON ACCURACY

Laser-engraved angle scales on this sliding-table jig (**3–11**) help guide you to exact miter cuts, taper cuts, crosscuts, and angle cuts. Time and time again, this jig has proved its worth in the *WOOD®* magazine shop where it was conceived, designed, and tested. Now, you can build one for your own

shop and raise your woodworking to a new level of accuracy. *See pages 72–74 for just a few of the jig's many uses.*

Start with the Jig Table

1 Cut the jig table (A) to the size listed in the Materials List on *page 70*. For an accurate cutting jig later, make sure the rectangular table you cut has perfectly square corners. (Due to its stability and strength, we used ¾" [18mm actual] Baltic birch plywood.)

2 To customize the jig table for your particular tablesaw, you'll need to properly locate the miter-gauge groove on the *bottom* side of the table (A). To do this, follow the three steps on **3–12**. Because one miter-gauge groove in a tablesaw is located farther from the blade on one side than the other, the slot in the jig's table (A) will not be centered.

3 Using a dado blade in your tablesaw, cut a ¹⁄₁₆"-deep dado on the *bottom* side of the table (A) to the same width as your tablesaw's miter-gauge groove centered between the lines, where marked in Steps 1 and 2 on **3–12**.

4 Cut the miter-gauge guide bar (B) to size from solid birch stock. (We found that although solid wood works, UHMW [ultra-high molecular weight] polyethylene slides easier in the groove than wood. Plus, polyethylene will not change in size with seasonal humidity changes.) Drill mounting holes in the guide, where shown on **3–13**. Screw the guide in place, making sure the screw heads don't protrude below the bottom surface of the guide.

5 Fit your tablesaw with a ¹³⁄₁₆" dado blade that's set to cut ⁷⁄₁₆" deep (¹⁄₁₆" deeper than the thickness of the metal mini channel). Set the fence on your tablesaw 3" from the edge of the dado blade. With a different edge against the fence for each pass, cut four dadoes in the *top surface* of the table (A), where indicated on **3–13**.

3–11.

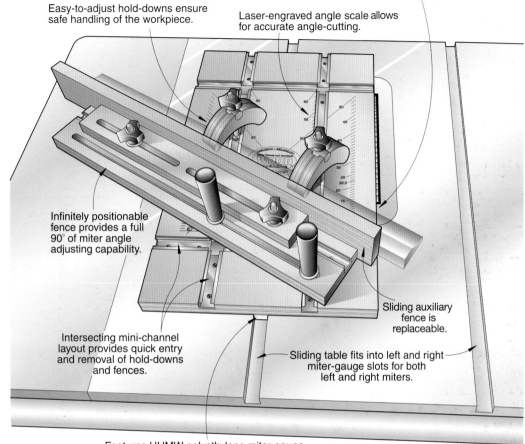

Edge of table aligns flush with blade for easy cutoff reference. This is especially useful when cutting odd-shaped parts and taper-cutting.

Easy-to-adjust hold-downs ensure safe handling of the workpiece.

Laser-engraved angle scale allows for accurate angle-cutting.

Infinitely positionable fence provides a full 90° of miter angle adjusting capability.

Sliding auxiliary fence is replaceable.

Intersecting mini-channel layout provides quick entry and removal of hold-downs and fences.

Sliding table fits into left and right miter-gauge slots for both left and right miters.

Features UHMW polyethylene miter-gauge guide for long-wearing durability.

3–12.

LOCATING THE GUIDE-BAR SLOT

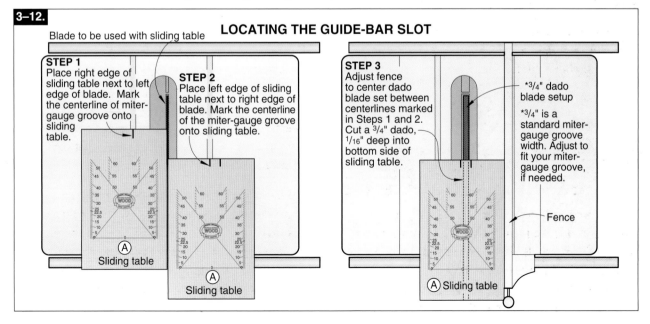

Blade to be used with sliding table

STEP 1
Place right edge of sliding table next to left edge of blade. Mark the centerline of miter-gauge groove onto sliding table.

STEP 2
Place left edge of sliding table next to right edge of blade. Mark the centerline of the miter-gauge groove onto sliding table.

Ⓐ Sliding table

Ⓐ Sliding table

STEP 3
Adjust fence to center dado blade set between centerlines marked in Steps 1 and 2. Cut a ³/₄" dado, ¹/₁₆" deep into bottom side of sliding table.

*³/₄" dado blade setup

*³/₄" is a standard miter-gauge groove width. Adjust to fit your miter-gauge groove, if needed.

Fence

Ⓐ Sliding table

6 Replace the dado set with the blade you normally use in your tablesaw. Place the jig in one of the miter-gauge slots and trim one edge of the jig with the blade. Mark the front end of the jig top. Transfer the jig into the other miter-gauge slot, and, with the marked end forward, trim the opposite edge of the jig.

Note: To save you the time and effort of having to use an adjustable triangle to position the fence to cut miters, we've supplied a source for a jig table (A) laser-engraved with angle scales. See how and where to order in the Materials List on page 70. If you'd rather use your own wood for the table, use an adjustable triangle to set the angle of cut, as shown in 3–24 on page 74.

Cut and Add the Metal Mini Channel

1 Measure the lengths, and use a hacksaw or bandsaw fitted with a metal-cutting blade to cut the 12 pieces of metal mini channel to length plus ⅛". (We used B-Line System B72 mini channel, available at most electrical supply outlets.)

2 To square the ends and make sure the corresponding pieces are identical in length, screw a wooden extension to your disc sander miter gauge, where shown in **3–15**. Then, mark three lines on the fence to indicate the three lengths of mini channel needed.

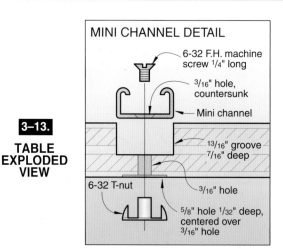

3–13.
TABLE EXPLODED VIEW

MINI CHANNEL DETAIL

6-32 F.H. machine screw ¼" long
³/₁₆" hole, countersunk
Mini channel
¹³/₁₆" groove ⁷/₁₆" deep
6-32 T-nut
³/₁₆" hole
⅝" hole ¹/₃₂" deep, centered over ³/₁₆" hole

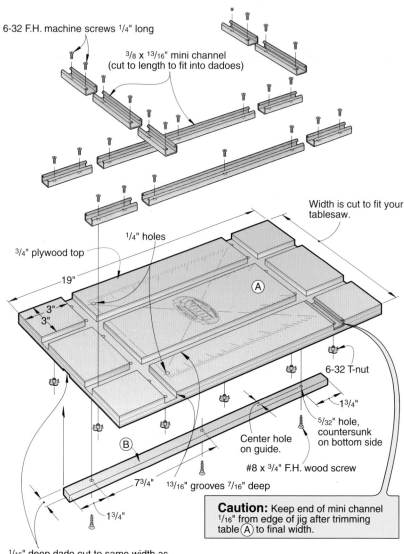

6-32 F.H. machine screws ¼" long
⅜ x ¹³/₁₆" mini channel (cut to length to fit into dadoes)
Width is cut to fit your tablesaw.
¼" holes
¾" plywood top
19"
3"
3"
A
6-32 T-nut
1¾"
⁵/₃₂" hole, countersunk on bottom side
Center hole on guide.
#8 x ¾" F.H. wood screw
B
7¾"
¹³/₁₆" grooves ⁷/₁₆" deep
1¾"
¹/₁₆" deep dado cut to same width as miter-gauge slot. Cut guide (B) to match width of your miter-gauge slot.

Caution: Keep end of mini channel ¹/₁₆" from edge of jig after trimming table (A) to final width.

Using a disc sander, sand one end of each piece of channel square. Then, sand the opposite end of each, pushing lightly on the already sanded end until it is flush with the previously marked lines.

Finally, sand the four short side pieces so they will be positioned ¹⁄₁₆" in from the edge of the table side, where noted on **3–13**. This prevents your tablesaw blade from coming in contact with the mini channel.

3 Fit your drill press with a ³⁄₁₆" bit, and attach a fence to the table so you can drill all the mounting holes centered in the top surface of the mini channel, where shown on the Mini Channel detail in **3–13** and in **3–28** on *page 75*. File off any burrs from the bottom side of the channel.

4 Remove the guide (B) from the bottom side of the table (A). Then, using double-faced tape, stick each piece of mini channel in its mating location.

3–14.
AUXILIARY ATTACHMENT

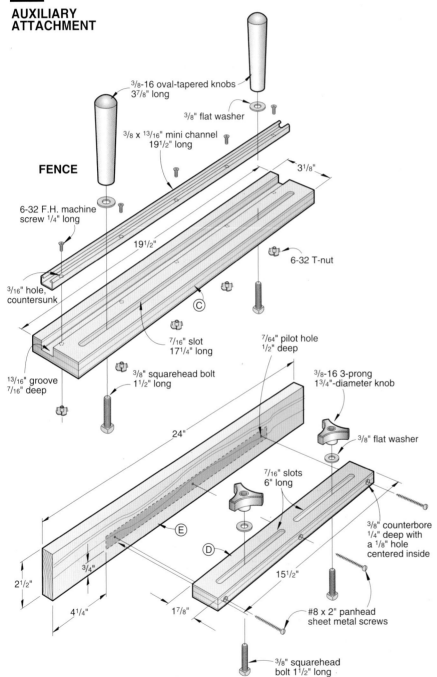

FENCE

3/8-16 oval-tapered knobs 3⁷⁄₈" long

3/8" flat washer

3/8 x 13/16" mini channel 19¹⁄₂" long

6-32 F.H. machine screw 1/4" long

19¹⁄₂"

3¹⁄₈"

6-32 T-nut

3/16" hole, countersunk

C

7/16" slot 17¹⁄₄" long

13/16" groove 7/16" deep

3/8" squarehead bolt 1¹⁄₂" long

24"

7/64" pilot hole 1/2" deep

3/8-16 3-prong 1³⁄₄"-diameter knob

3/8" flat washer

7/16" slots 6" long

E

D

3/8" counterbore 1/4" deep with a 1/8" hole centered inside

2¹⁄₂"

3/4"

15¹⁄₂"

4¹⁄₄"

1⁷⁄₈"

#8 x 2" panhead sheet metal screws

3/8" squarehead bolt 1¹⁄₂" long

3–15.
Marked lines on a miter-gauge extension allow you to sand the mini-channel to the exact lengths needed.

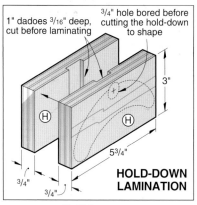

1" dadoes ³/₁₆" deep, cut before laminating

³/₄" hole bored before cutting the hold-down to shape

3"

5³/₄"

³/₄"

³/₄"

H H

HOLD-DOWN LAMINATION

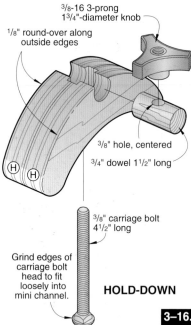

³/₈-16 3-prong 1³/₄"-diameter knob

¹/₈" round-over along outside edges

³/₈" hole, centered

³/₄" dowel 1¹/₂" long

H H

³/₈" carriage bolt 4¹/₂" long

Grind edges of carriage bolt head to fit loosely into mini channel.

HOLD-DOWN

3–16.

Materials List for Universal Jig With Right-On Accuracy

PART	T	W	L	Matl.	Qty.
A table	³/₄"	12"	19"	BP	1
B guide	³/₈"	**	19	B	1
C fence	³/₄"	3¹/₈"	19¹/₂"	BP	1
D base	³/₄"	1⁷/₈"	15¹/₂"	BP	1
E upright	³/₄"	2¹/₂"	24"	B	1
F stopblock	³/₄"	1⁵/₈"	1⁵/₈"	BP	1
G stopblock	¹/₂"	1⁵/₁₆"	1⁵/₈"	B	1
H hold-down blanks	³/₄"	3"	5³/₄"	BP	4
I holder	1¹/₂"	2"	3"	P	1

**Width depends on the width of your auxiliary tablesaw's miter-gauge groove.

Materials Key: BP = Baltic birch plywood; B = birch; P = pine or fir.

Supplies: Thirteen pieces of ³/₈" x 1³/₁₆" mini channel; thirty-one 6–32 T-nuts and mating 6–32 flathead machine screws, ¹/₄" long; three #8 x ³/₄" flathead wood screws; two ³/₈–16 oval-tapered knobs, 3 ⁷/₈" long; five ³/₈" flat washers; five ³/₈" squarehead bolts, 1¹/₂" long; five ³/₈–16 three-pronged plastic knobs, 1³/₄" diameter; three #8 x 2"panhead sheet metal screws; two #8 x 1" flathead brass wood screws; ³/₄" dowel stock; two ³/₈" carriage bolts, 4¹/₂" long; four ¹/₂"-diameter magnets ¹/₄" thick; two pieces of ¹/₄" steel rod 5" long; clear finish.

Buying Guide
Laser-engraved jig table, part A (item #TSJ-PANEL) with dadoed mini-channel grooves is available from Schlabaugh and Sons Woodworking for under $30. Visit their Website www.schasons.com or call 800-346-9663 or 319-656-2374.

Using the same ³/₁₆" bit in your drill press, use the holes in the mini channel as guides to drill ³/₁₆" holes through the table (A).

5 Working from the bottom side of the jig table, drill a ⁵/₈" counterbore ¹/₃₂" deep for each T-nut. Test the fit; the T-nuts must not come in contact with the metal top of your tablesaw when the jig slides back and forth.

6 One at a time, remove the channel and attach a piece of masking tape to the bottom side of each piece. For ease in relocating the channel later, number the tape on the channel and mark the mating number on the dado from where the channel was removed.

7 Lightly sand the table and apply a couple of light coats of finish. (We used polyurethane.)

8 Tap all the T-nuts into place in the bottom of the table (A).

9 Using the numbers as guides, reposition the mini channel in their mating locations in the jig table. Fit a countersink bit (we used a Weldon ³/₈"-diameter

countersink bit) into your drill press, and countersink each hole in the mini channel so the top surface of a 6–32 flathead machine screw will seat in the countersink and the top of the screw will be flush with the top of the channel. For the stops and fences to slide smoothly in the channel later, the tops of the screws must not protrude. The countersink bit will also machine the tops of the T-nuts so they don't protrude through the mini channel.

10 Remove the pieces of mini channel from the table. Use the countersink bit to slightly machine the tops of the T-nuts a bit more. This creates a slight gap between the T-nuts and mini channel so the T-nut will be pulled tightly into the bottom of the table when securing the mini channel in place. Screw the channel in place.

11 Reattach the guide bar (B) to the bottom of the table.

Make the Fence and the Auxiliary Attachment

1 Cut the fence (C) to the size listed in the Materials List and shown on **3–29**, on *page 75*.

2 Mark and drill a pair of $\frac{7}{16}$" holes through the fence (C), where shown on **3–29**. Draw lines to connect the holes, and cut the waste between the holes to form a $\frac{7}{16}$" slot.

3 Cut a $\frac{13}{16}$" dado $\frac{7}{16}$" deep in the top surface of the fence, where dimensioned on **3–29** and shown on the Fence drawing in **3–14,** on *page 69*.

4 Cut the mini channel for the fence (C) to length, and drill and countersink the mounting holes as you did earlier for the table (A). Finish-sand the fence, and screw the channel in place.

5 Cut the auxiliary attachment base (D) and upright (E) to size.

6 Mark the locations and form $\frac{7}{16}$" slots in the base (D), where shown on **3–27**, on *page 75*.

7 Mark the centerpoints, drill three $\frac{1}{8}$" counterbored mounting holes, and screw the upright (E) to the base (D), where shown on the Auxiliary Attachment in **3–14** and **3–27** on *page 75*. Do not glue D to E, because you'll need to replace E after you've cut through it numerous times.

For Cuts of Equal Length, Add a Stopblock

1 Cut the stopblock horizontal piece (F) and vertical piece (G) to size. See the Stopblock drawing in **3–17** for reference.

2 Drill a $\frac{3}{8}$" hole in F, where dimensioned on the Stopblock drawing in **3–17**. Then, drill a pair of countersunk mounting holes through G and into F. Screw the pieces together.

3 Attach a knob to F, where shown in the Stopblock drawing in **3–17**.

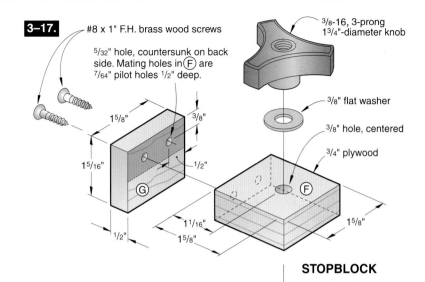

3–17.

#8 x 1" F.H. brass wood screws

$\frac{5}{32}$" hole, countersunk on back side. Mating holes in Ⓕ are $\frac{7}{64}$" pilot holes $\frac{1}{2}$" deep.

$\frac{3}{8}$-16, 3-prong $1\frac{3}{4}$"-diameter knob

$\frac{3}{8}$" flat washer

$\frac{3}{8}$" hole, centered

$\frac{3}{4}$" plywood

$1\frac{5}{8}$"

$\frac{3}{8}$"

$1\frac{5}{16}$"

$\frac{1}{2}$"

Ⓖ

$\frac{1}{2}$"

$1\frac{1}{16}$"

$1\frac{5}{8}$"

Ⓕ

$1\frac{5}{8}$"

STOPBLOCK

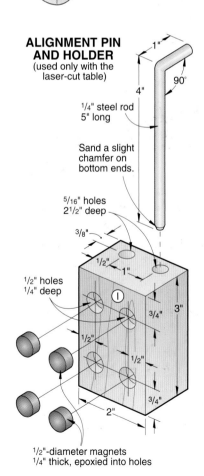

$\frac{3}{8}$" squarehead bolt $1\frac{1}{2}$" long

ALIGNMENT PIN AND HOLDER
(used only with the laser-cut table)

1"

90°

4"

$\frac{1}{4}$" steel rod 5" long

Sand a slight chamfer on bottom ends.

$\frac{5}{16}$" holes $2\frac{1}{2}$" deep

$\frac{3}{8}$"

$\frac{1}{2}$"

1"

$\frac{1}{2}$" holes $\frac{1}{4}$" deep

Ⓘ

$\frac{3}{4}$"

3"

$\frac{1}{2}$"

$\frac{1}{2}$"

$\frac{3}{4}$"

2"

$\frac{1}{2}$"-diameter magnets $\frac{1}{4}$" thick, epoxied into holes

Finally, Add a Pair of Hold-Downs and Pivot Pins

1 Cut four hold-down blanks (H) to 3 × 5¾" from ¾" plywood. Cut a 1" dado ³⁄₁₆" deep in each piece, where shown on the Hold-Down Lamination drawing in **3–16**. With the edges and ends aligned, glue and clamp two pieces together to form each hold-down. Later, transfer the patterns in **3–30** on *page 76*, bore a ¾" hole where indicated, and band-saw the hold-downs to shape.

2 Drill and cut a pair of the ¾" dowels shown on the Hold-Down drawing in **3–16**, and assemble the hold-downs in the configuration shown on the drawing.

3 If you don't plan to buy the laser-engraved table (A), skip to the next section. Crosscut and bend two pieces of ¼" steel rod 5" long to the shape shown on the Alignment Pin and Holder drawing in **3–17**. You'll use the alignment pins for aligning the fence on the laser-engraved table later.

4 Build the alignment-pin holder shown on the drawing. To keep the holder close at hand, drill counterbores and epoxy four magnets in place for sticking the holder to your tablesaw cabinet.

Sand and Apply a Finish

1 Finish-sand the fence, auxiliary attachment, stop, hold-downs, and alignment-pin holder.

2 For your plywood table (A), consider marking commonly used angles on the top surface. Find the angles with an adjustable triangle, as shown in **3–24** (*page 74*).

3 Add a clear finish to all wood parts. (To prevent the working surfaces from becoming too slick, we prefer a polyurethane finish.)

4 If you used solid stock for your guide (B), apply a bit of paraffin to the guide for easier sliding in the tablesaw groove.

How to Use the Jig

To cut angled pieces to shape, mark the cutlines on the wood. Align a marked cutline with the edge of the jig table. Position the fence and hold-downs to hold the piece steady. Make the cut, as shown in **3–18**. The fence adjusts easily for making the adjoining angle cut, as shown in **3–19**.

You can taper-cut table legs and other projects by marking the cutline on the workpiece and aligning the marked cutline with the outside edge of the jig table. Then, as shown in **3–20**, position the fence against the workpiece, add the stop and hold-downs, and make the cut. With this setup, you can cut numerous pieces exactly the same.

For making accurate 90° crosscuts, use the pivot pins to accurately align the fence on the laser-cut table, as shown in **3–21**. Use the jig and fence for crosscutting, as shown in **3–22**. For repetitive cuts, secure the stopblock to the fence to ensure consistent lengths from piece to piece.

To set the angle using the laser-engraved table, fit one pivot pin in place and pivot the fence against the pin. Align the fence with the laser-marked angle, as shown in **3–23**. If you use your own stock for the jig table, use an adjustable triangle and a piece of scrap stock held against the edge of the jig table to correctly angle the fence, as shown in **3–24**.

Accurately cut both left and right miters with this jig, as shown in **3–25** and **3–26**. Note the use of the auxiliary attachment. For crown molding and picture frame material, you'll need to move the jig to the opposite side of the blade, as shown in **3–26**.

Hold-Down Patterns

When using our jig, it became apparent that different-size hold-downs were necessary for the many sizes of workpieces we needed to secure to the sliding table and fence. Use the full-size patterns shown in **3–30**, on *page 76,* to make your own hold-downs.

3-18.

3-19.

3-20.

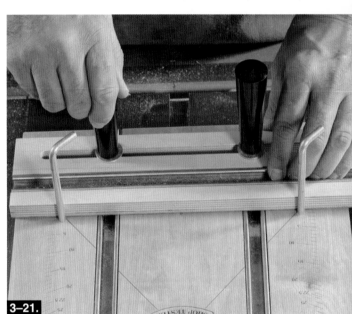
3-21.

3-22.

3-23.

3–24.

3–25.

3–26.

3–27.

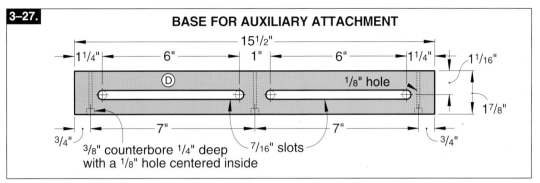

BASE FOR AUXILIARY ATTACHMENT

15¹/₂"
1¹/₄" · 6" · 1" · 6" · 1¹/₄" · 1¹/₁₆"
(D) · ¹/₈" hole · 1⁷/₈"
7" · 7"
³/₄" · ³/₄"
³/₈" counterbore ¹/₄" deep with a ¹/₈" hole centered inside
⁷/₁₆" slots

3–28.

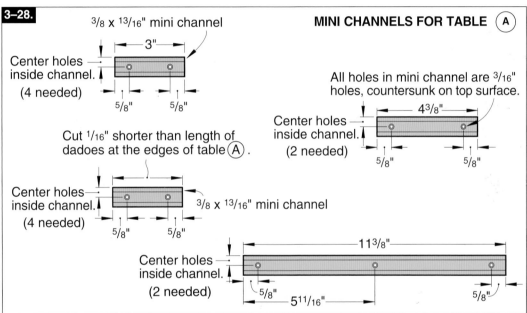

MINI CHANNELS FOR TABLE (A)

³/₈ x ¹³/₁₆" mini channel
3"
Center holes inside channel. (4 needed)
⁵/₈" · ⁵/₈"

All holes in mini channel are ³/₁₆" holes, countersunk on top surface.
Center holes inside channel. (2 needed)
4³/₈"
⁵/₈" · ⁵/₈"

Cut ¹/₁₆" shorter than length of dadoes at the edges of table (A).

Center holes inside channel. (4 needed)
³/₈ x ¹³/₁₆" mini channel
⁵/₈" · ⁵/₈"

Center holes inside channel. (2 needed)
11³/₈"
⁵/₈" · 5¹¹/₁₆" · ⁵/₈"

3–29.

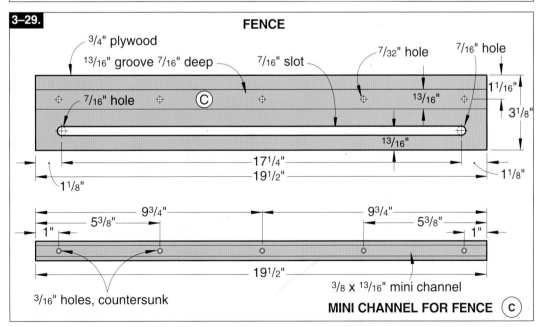

FENCE

³/₄" plywood
¹³/₁₆" groove ⁷/₁₆" deep
⁷/₁₆" slot
⁷/₃₂" hole
⁷/₁₆" hole
1¹/₁₆"
⁷/₁₆" hole
(C)
¹³/₁₆"
3¹/₈"
¹³/₁₆"
17¹/₄"
19¹/₂"
1¹/₈"
1¹/₈"

9³/₄" · 9³/₄"
5³/₈" · 5³/₈"
1" · 1"
19¹/₂"
³/₁₆" holes, countersunk
³/₈ x ¹³/₁₆" mini channel

MINI CHANNEL FOR FENCE (C)

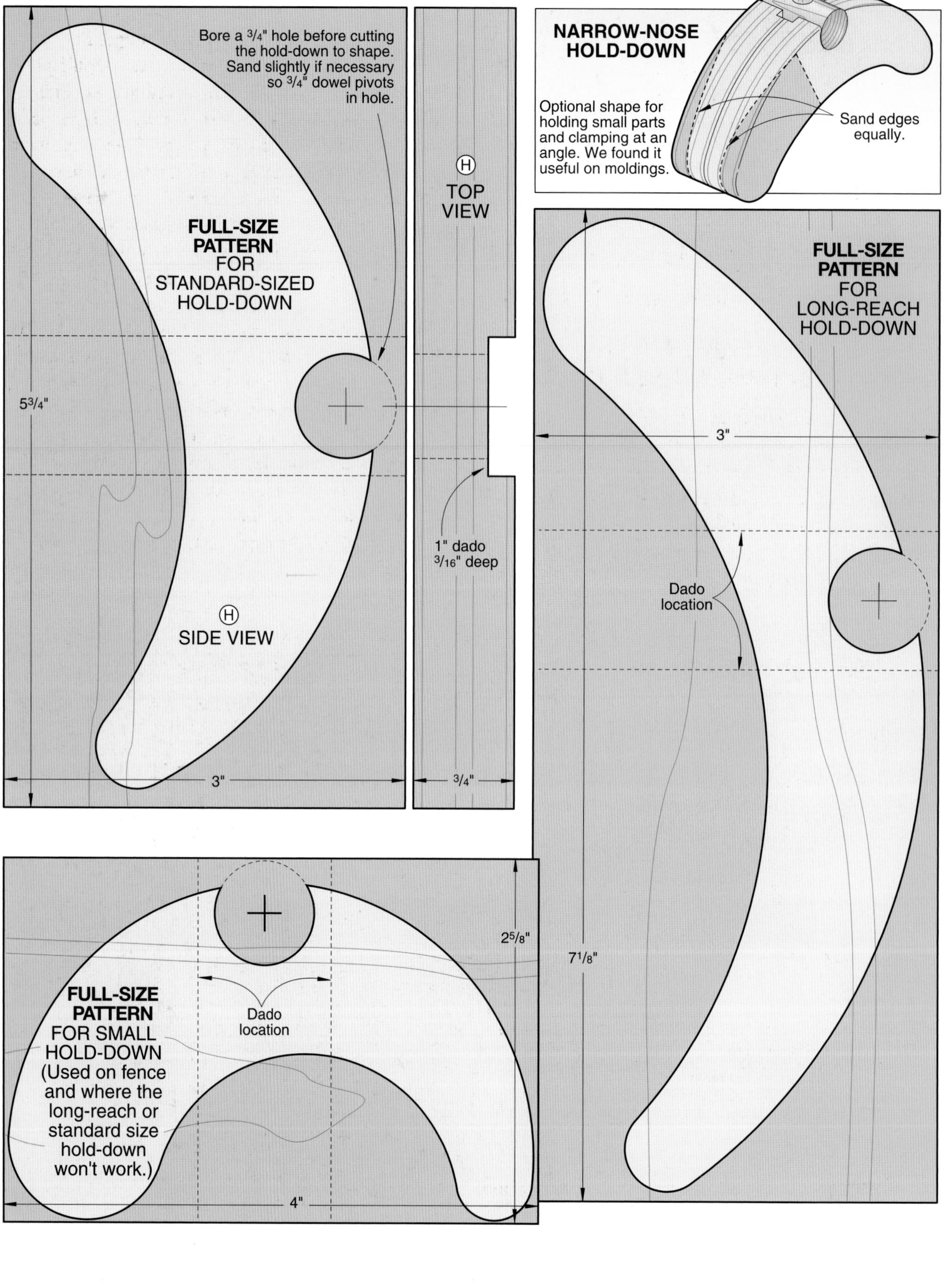

Bore a ³/₄" hole before cutting
the hold-down to shape.
Sand slightly if necessary
so ³/₄" dowel pivots
in hole.

**NARROW-NOSE
HOLD-DOWN**

Optional shape for
holding small parts
and clamping at an
angle. We found it
useful on moldings.

Sand edges
equally.

**FULL-SIZE
PATTERN**
FOR
STANDARD-SIZED
HOLD-DOWN

**TOP
VIEW**

Ⓗ

5³/₄"

Ⓗ
SIDE VIEW

1" dado
³/₁₆" deep

3"

³/₄"

**FULL-SIZE
PATTERN**
FOR
LONG-REACH
HOLD-DOWN

3"

Dado
location

**FULL-SIZE
PATTERN**
FOR SMALL
HOLD-DOWN
(Used on fence
and where the
long-reach or
standard size
hold-down
won't work.)

Dado
location

2⁵/₈"

7¹/₈"

4"

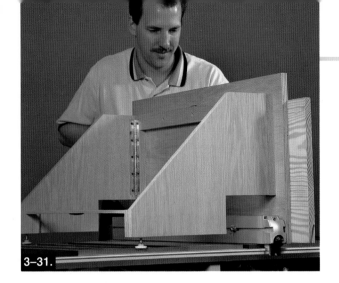

3–31.

TEXAS-SIZE TABLESAW FENCE

You can stand workpieces upright and cut their edges on your tablesaw using this auxiliary tablesaw fence (**3–31** and **3–32**), designed by *WOOD*® magazine reader Joe Xaver. The jig bolts temporarily to your saw's existing fence to let you make these cuts safely and accurately, and folding supports make for flat storage.

First, measure your saw to ensure proper fit. Then examine the saw's existing fence to see if drilling it for machine screws will interfere with its operation; adjust the location if necessary. For webbed extension wings, measure between the centers of the webs at the front and rear of the extensions. Make the removable spreader this length, and add 1½" to find the length of the crossmember. (The dimensions shown are for a table that is 26½" between the centers of the front and rear webs.) For saws with solid extension wings, shorten the dimensions for those pieces by 4".

Armed with that information, build the auxiliary fence, as shown in **3–32**. Drill ⁵⁄₁₆" holes in your fence to match the location of the T-nuts, and bolt the tall fence to your saw's fence, as shown in **3–31**.

Adjust the nylon glides so the tall fence is perpendicular to your saw's table top. When done with a job, unbolt the unit from your fence, pop out the spreader (store it on the top edge of the crossmember), fold up the legs, and hang the jig on a wall.

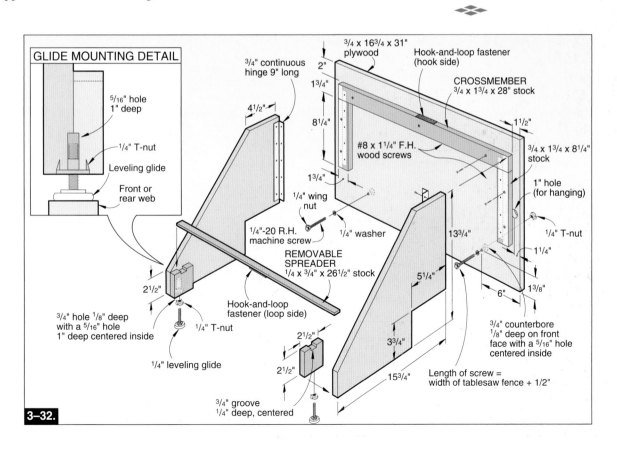

3–32.

3–33.

Positioned to center the workpiece over the workpiece over the dado blade, the jig is the perfect setup for machining bridle joints or open mortises and the mating tenons.

RIP-FENCE SADDLE

Build this auxiliary wood fence and mating saddle to bevel-cut such things as post caps, or build it for supporting stiles and other workpieces, as shown in **3–33**. Use one hand to push the saddle and workpiece across the blade, and your other hand to keep the saddle riding firmly on the auxiliary fence. Wax the mating pieces if necessary for easy sliding.

Note: Our auxiliary fence is screwed securely to our metal tablesaw rip fence, with the top edge of the fence sitting 1" above the top edge of the metal fence. The auxiliary fence must be 90° to the saw table. Size your wood fence so the saddle rides smoothly, without free play, along the top edge of the auxiliary fence.

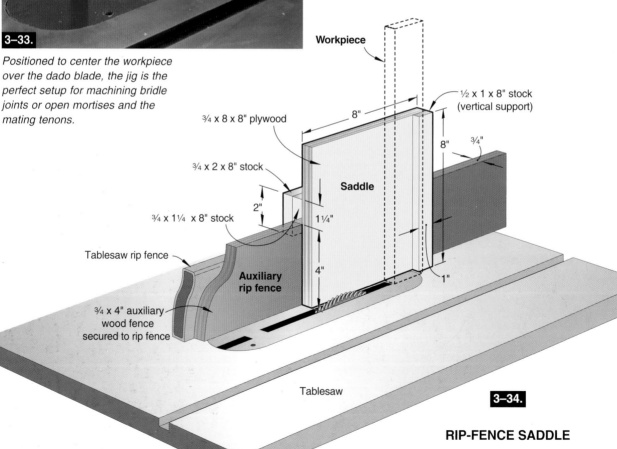

3–34.

RIP-FENCE SADDLE

BEVEL-CUTTING JIG FOR SMALL PARTS

3–35.

3–37.

As you probably already know, machining small parts—especially sawing them—can be both tricky and dangerous. Here's a novel way to cut bevels on small parts without getting your hands in harm's way.

The toggle clamp attached to the jig, shown in **3–35,** securely captures the small part as the jig slides forward along the rip fence. See **3–36** for how the jig goes together. Make one of these for each degree bevel you want to cut.

TENONING JIG

Mortise-and-tenon joinery is strong. And, this easy-to-make tablesaw jig (**3–37** and **3–38**) is the accessory that makes it all possible.

Start with the Base and Sliding Table

1 Cut two pieces of ¾" plywood for the base (A) and two pieces for the sliding table (B) to the sizes listed in the Materials List on *page 82,* plus ½" in length and width. (Due to its stability and strength, we used ¾" [18mm actual] Baltic birch plywood.)

2 With the edges and ends flush, glue and clamp the two base pieces together face-to-face. Repeat with the two remaining pieces to form the sliding table. Later, remove the clamps and cut the base (A) and sliding table (B) to the finished sizes listed in the Materials List, on *page 82.*

3–36.
SMALL-PART BEVEL JIG

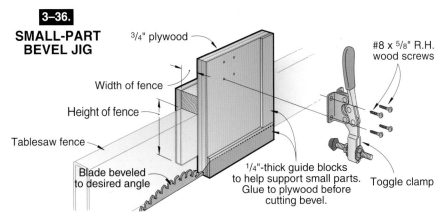

¾" plywood

#8 x ⅝" R.H. wood screws

Width of fence

Height of fence

Tablesaw fence

Blade beveled to desired angle

¼"-thick guide blocks to help support small parts. Glue to plywood before cutting bevel.

Toggle clamp

3 Measure the exact width of the miter-gauge groove in your tablesaw. Cut the base guide bar (C) to size. Use solid maple or birch. The guide should slide in the groove without slop. Set it aside; you'll add it later.

4 Using a dado head in your tablesaw, cut a dado the width of your guide bar (C) and ⅛" deep on the bottom side of the base (A), where dimensioned on **3–39** and **3–40**. The guide bar (C) will fit into this dado later.

5 Cut a ¾" dado ⅛" deep on the top side of the base (A) and a mating ¾" dado ³⁄₁₆" deep on the bottom side of the sliding table (B), where shown on **3–40**. Later,

you'll screw the guide bar (D) into the dado in the top of the base. And when assembled, the sliding table (B) will slide on the top, exposed portion of this guide bar.

6 Cut a ¾" dado ¹⁄₁₆" deep in the top of the base (A) to house the 6" metal rule.

7 Cut the second guide bar (D) so it fits snugly in the top dado in the base, and slides smoothly in the ¾" dado on the bottom of the sliding table. Set this guide aside also.

8 Mark the centerpoints, where dimensioned on **3–40**, and drill the holes for the magnets in the dado in the top of the base (A). Measure your magnets

before drilling; they may vary in size. You want the magnets to sit just a hair below the surface of the dado.

9 Mark the centerpoint, and drill a ⅞" hole ½" deep on the bottom side of the base (A). Then, drill a ⅜" hole centered in the ⅞" hole and through the base for the ⅜" carriage bolt. Check the fit to make sure the bottom of the carriage bolt doesn't protrude.

10 Mark a pair of center-points, drill a ⁷⁄₁₆" hole at each point, and cut between the holes with a scroll-saw or jigsaw to form the ⁷⁄₁₆"-wide slot in the sliding table (B), where dimensioned on **3–40**.

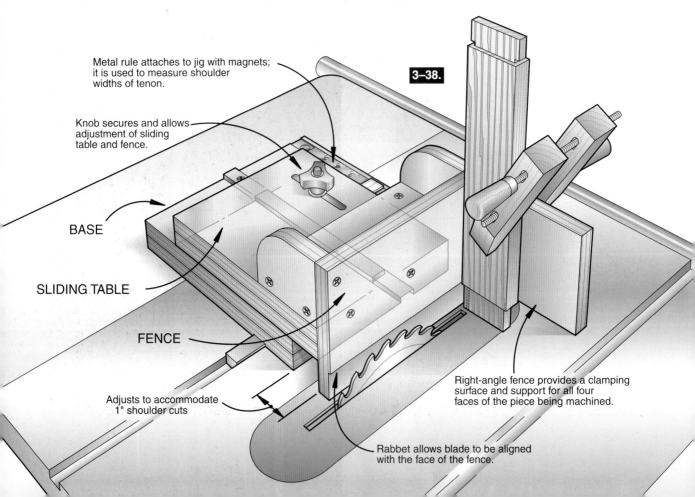

Metal rule attaches to jig with magnets; it is used to measure shoulder widths of tenon.

Knob secures and allows adjustment of sliding table and fence.

3–38.

BASE

SLIDING TABLE

FENCE

Adjusts to accommodate 1" shoulder cuts

Right-angle fence provides a clamping surface and support for all four faces of the piece being machined.

Rabbet allows blade to be aligned with the face of the fence.

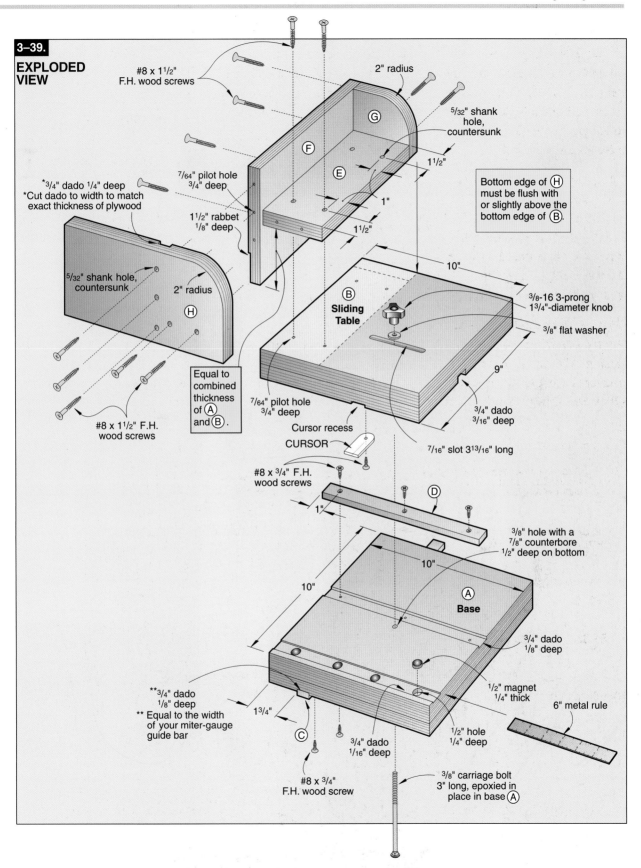

3–39.

EXPLODED VIEW

#8 x 1¹/₂" F.H. wood screws

2" radius

⁵/₃₂" shank hole, countersunk

1¹/₂"

*³/₄" dado ¹/₄" deep
*Cut dado to width to match exact thickness of plywood

⁷/₆₄" pilot hole ³/₄" deep

1¹/₂" rabbet ¹/₈" deep

1"

1¹/₂"

Bottom edge of Ⓗ must be flush with or slightly above the bottom edge of Ⓑ.

⁵/₃₂" shank hole, countersunk

2" radius

Ⓖ

Ⓕ

Ⓔ

Ⓗ

Equal to combined thickness of Ⓐ and Ⓑ.

#8 x 1¹/₂" F.H. wood screws

⁷/₆₄" pilot hole ³/₄" deep

Cursor recess

CURSOR

#8 x ³/₄" F.H. wood screws

10"

Ⓑ
Sliding Table

³/₈-16 3-prong 1³/₄"-diameter knob

³/₈" flat washer

9"

³/₄" dado ³/₁₆" deep

⁷/₁₆" slot 3¹³/₁₆" long

Ⓓ

1"

³/₈" hole with a ⁷/₈" counterbore ¹/₂" deep on bottom

10"

10"

Ⓐ
Base

³/₄" dado ¹/₈" deep

¹/₂" magnet ¹/₄" thick

6" metal rule

**³/₄" dado ¹/₈" deep
** Equal to the width of your miter-gauge guide bar

1³/₄"

Ⓒ

³/₄" dado ¹/₁₆" deep

¹/₂" hole ¹/₄" deep

#8 x ³/₄" F.H. wood screw

³/₈" carriage bolt 3" long, epoxied in place in base Ⓐ

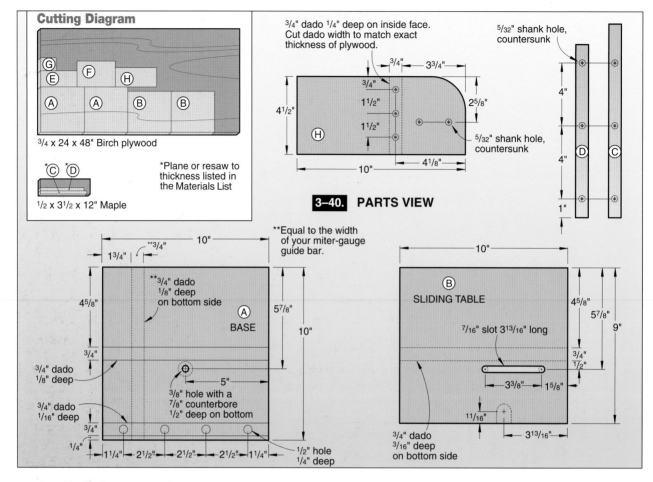

Cutting Diagram

3/4 x 24 x 48" Birch plywood

*Plane or resaw to thickness listed in the Materials List

1/2 x 3 1/2 x 12" Maple

3/4" dado 1/4" deep on inside face. Cut dado width to match exact thickness of plywood.

5/32" shank hole, countersunk

3–40. PARTS VIEW

**Equal to the width of your miter-gauge guide bar.

BASE

SLIDING TABLE

Materials List for Tenoning Jig

PART	FINISHED SIZE				
	T	W	L	MTL.	QTY.
A* base	1 1/2"	10"	10"	LBP	1
B* sliding table	1 1/2"	9"	10"	LBP	1
C guide bar	7/16"	3/4"	11"	M	1
D guide bar	5/16"	3/4"	10"	M	1
E* horiz. support	3/4"	3 3/4"	8 1/4"	BP	1
F fence	3/4"	6"	9 1/4"	BP	1
G radiused end	3/4"	3"	3 3/4"	BP	1
H dadoed end	3/4"	4 1/2"	10"	BP	1

*Initially cut parts oversized. Trim to finished size according to the instructions.

Materials Key: LBP = laminated Baltic birch plywood; BP = Baltic birch plywood; M = maple.
Supplies: #8 x 3/4" flathead wood screws (7); #8 x 1 1/2 flathead wood screws (15); 1/2" magnets 1/4" thick (4); 6" metal rule; 3/8" carriage bolt 3" long with mating washer and plastic knob; 1/8 x 4 x 4" acrylic for cursor; polyethylene for the guide bar (C); epoxy; clear finish.

11 Drill countersunk mounting holes in the guides (C, D), where dimensioned on **3–40**. Screw the guides in place, making sure the screw heads don't protrude.

Form the Cursor Next

1 Cut a piece of 1/8" clear acrylic to 4 x 4". Chuck a 7/8" Forstner bit into your drill press, and position the bit over the acrylic, roughly where shown on **3–42**. Clamp the acrylic securely to your drill-press table.

2 Start the drill press, and slowly lower the bit until the outside edge and centerpoint of the bit just barely score the acrylic.

3 Using a small square and a crafts knife held sideways (it scores better this way), score the three lines on the bottom side of the acrylic, as shown in **3–41** and where shown on **3–42**.

Drag the blade of a craft knife sideways to scribe the centerline and two cutlines on the acrylic cursor blank.

4 To make the centerline on the cursor more visible, use a felt-tipped marker to highlight the middle scribed line. Wipe any excess marker off the surface of the cursor.

5 Using your bandsaw fitted with a ⅛" blade or a scrollsaw with a #10 blade, cut the cursor to shape. Sand the edges smooth.

6 Working from the bottom side of the cursor, drill a ⁵⁄₃₂" countersunk shank hole centered over the bit centerpoint, where shown on **3–43**.

7 Using **3–44** for reference, form the cursor recess on the bottom side of the sliding table (B).

Add the Workpiece Support

*Note: Plywoods vary in thickness. All dimensions are based on plywood measuring exactly ¾" thick. See the tinted boxes on **3–39** before locating the horizontal support (E) against the fence (F).*

1 Cut the horizontal support (E) to size plus ½" in length. Then, cut the fence (F), radiused end (G), and dadoed end (H).

2 Cut the rabbet along the outside face of the fence (F), where shown on the **3–39**.

3 Cut a ¼"-deep dado in the inside face of H to the same exact thickness as your plywood.

4 Mark the centerpoints, and drill holes in H, where dimensioned on the **3–40**. Clamp the end (H) to the fence (F) and drive the screws. Drill the holes,

and screw the opposite end (G) to the fence. Measure the distance between the ends (G, H), and cut the horizontal support (E) to final length and screw it in place, making sure parts G and H meet at right angles to part E.

5 Clamp the support (E/F/G/H) to the sliding table (B), and screw the two assemblies together, keeping the outside face of F square to the sliding table.

Finishing and Final Assembly

1 Finish-sand all the pieces and seal with polyurethane.

2 Cut a piece of wood to ¾ × 1¹⁄₁₆ × 12". Then, put a drop of epoxy in each magnet hole, fit the magnets into the holes, and wipe off any excess epoxy. As shown in **3–45**, position the strip of wood over the magnets, and use it as a clamping bar to hold the magnets in place until the epoxy cures. Later, remove the clamping block and clamps.

3–42.
FORMING THE CURSOR

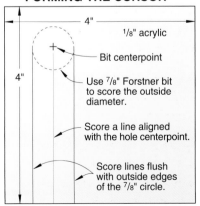

- 4"
- ⅛" acrylic
- 4"
- Bit centerpoint
- Use ⅞" Forstner bit to score the outside diameter.
- Score a line aligned with the hole centerpoint.
- Score lines flush with outside edges of the ⅞" circle.

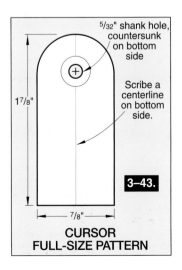

- ⁵⁄₃₂" shank hole, countersunk on bottom side
- 1⅞"
- Scribe a centerline on bottom side.
- **3–43.**
- ⅞"

CURSOR FULL-SIZE PATTERN

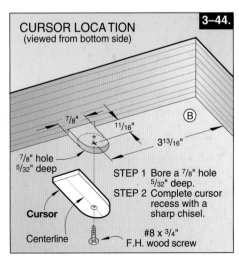

3–44.
CURSOR LOCATION
(viewed from bottom side)

(B)

- ⅞"
- 1¹⁄₁₆"
- 3¹³⁄₁₆"
- ⅞" hole ⁵⁄₃₂" deep
- **Cursor**
- **Centerline**
- STEP 1 Bore a ⅞" hole ⁵⁄₃₂" deep.
- STEP 2 Complete cursor recess with a sharp chisel.
- #8 × ¾" F.H. wood screw

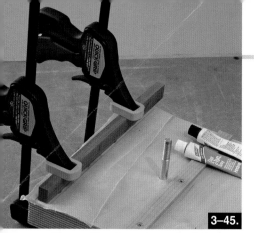

3–45.

Use a strip of wood, a pair of clamps, and waxed paper to hold the magnets in place until the epoxy cures.

3 Screw the acrylic cursor in place in the recess on the bottom of the sliding table (B).

4 Position the sliding table (B) on the base (A). Slide a ⅜" carriage bolt through the ⅜" hole in the base (A) and through the ⁷⁄₁₆" slot in the sliding table (B). Slide the two assemblies back and forth to check the fit, and then epoxy the carriage bolt in place. Attach a washer and plastic knob onto the bolt, where shown on **3–39**. To prevent the jig from possibly rocking on the saw table, make sure that the bottom edge of the fence (F) is flush with the bottom surface of the base (A). If F is higher, you may encounter a bit of rocking when the sliding table/fence is slid away from the base when cutting tenons. This can result in poorly cut tenons.

5 Lay the metal 6" rule in place on the magnets in the shallow dado.

THIN-STRIP RIPPING JIG

S ometimes you need to rip several thin strips of wood to equal thickness to serve as edging, veneer, or bending stock, but slicing off thin stock on the fence side of the blade could prove unsafe. That's because it becomes awkward to use your blade guard and pushstick when you cut close to the fence. The solution: Run the wide portion of your workpiece between the fence and blade, cutting the strips on the side of the blade opposite the fence. You could accomplish this by measuring for each cut, but that's tedious and inaccurate. This thin-strip ripping jig (**3–46**

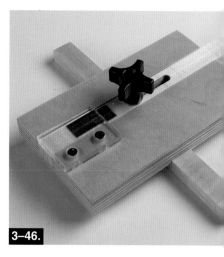

3–46.

and **3–47**) does the job safely, accurately, and quickly.

We used Baltic birch plywood and hard maple for the wood parts. If you prefer, you can substitute medium-density fiberboard (MDF) for plywood and another dense hardwood for maple.

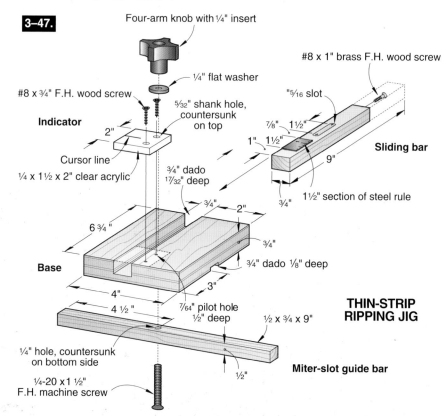

3–47.

Four-arm knob with ¼" insert

¼" flat washer

#8 x ¾" F.H. wood screw

5⁄32" shank hole, countersunk on top

Indicator

2"

Cursor line

¼ x 1½ x 2" clear acrylic

¾" dado 17⁄32" deep

#8 x 1" brass F.H. wood screw

"5⁄16 slot

⅞" 1½"

1" 1½"

9"

Sliding bar

1½" section of steel rule

¾" 2"

6¾"

¾"

¾"

Base

¾" dado ⅛" deep

3"

4"

4½"

7⁄64" pilot hole ½" deep

½ x ¾ x 9"

THIN-STRIP RIPPING JIG

¼" hole, countersunk on bottom side

½"

Miter-slot guide bar

¼-20 x 1½" F.H. machine screw

First, Build the Jig

1 Cut a piece of ¾" plywood to the dimensions shown for the base on **3–47**. Cut a dado on the bottom side of the base for the guide bar, where shown. Now, cut the ¾" dado on the top side of the base for the sliding bar.

2 Cut two pieces of maple to size for the miter-slot guide bar (adjust the dimensions shown if necessary to fit your tablesaw's slots) and the sliding bar. Center the miter-slot guide bar in the bottom dado, and glue it in place. Drill a pair of ⁵⁄₁₆" holes in the sliding bar, where shown; scroll-saw the material between them; and smooth the inside of the slot with a file.

3 Set the jig in your tablesaw's left miter-gauge slot. Place the sliding bar in the dado with its left end flush with the base. Slide the jig forward, and mark the point where a left-leaning

sawblade tooth touches the bar. Make a second mark ½" closer to the base. Remove the bar, and crosscut it at the second mark.

4 Drill a ⁷⁄₆₄" pilot hole in the sliding bar, centered on the end you just cut. Drive a brass screw halfway into the wood. (We used brass to avoid any chance of damaging a tablesaw blade.) You'll turn this screw in or out to fine-tune your jig's basic "zero" setting, or to adjust it for a blade of different thickness or with a different tooth set.

5 From the bottom side of the assembly, drill and countersink a ¼" hole through the miter-slot guide bar and base for the machine screw that holds the plastic knob. Sand all of the wood parts to 180 grit, and apply three coats of clear finish.

6 Make a mark 1" from the left end of the sliding bar. Cut the first 1½" from an inexpensive

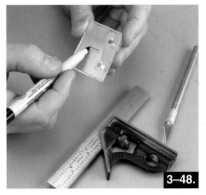

3–48.

To make a cursor, scribe a line across the middle of the acrylic indicator with a sharp knife and a combination square. Color the scribed line with a permanent marker. Wipe off the excess ink with a cloth or paper towel, leaving a fine line.

steel rule, align its left end with the mark, and attach it to the sliding bar with epoxy.

7 Cut a piece of ¼" acrylic plastic to the dimensions shown for the indicator. Drill and countersink the two mounting holes, and scribe and mark a cursor line, as described in the caption for **3–48**. Attach the indicator to the base, and add the knob.

Size your thin-strip ripping jig to suit your tablesaw, so that a 1" screw in the guide bar can contact the blade. Install a zero-clearance throat plate to prevent the sawn strip from falling into the saw.

3–49.

Remove the jig before making the cut so the workpiece doesn't bind between the rip fence and the screw head. Replace the jig in the slot without making any adjustments to set up the next cut.

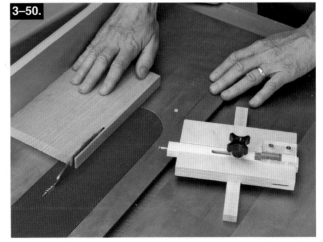

3–50.

Now, Cut Some Strips

To cut a thin strip with the jig, place its guide bar in the left-hand miter gauge slot on your tablesaw. Loosen the knob, set the cursor to zero (the bottom end of the rule), and retighten the knob. Slide the jig so the brass screw head is beside the saw blade. Turn the screw in or out with a screwdriver until the head lightly contacts a left-leaning tooth. Pull the jig toward you, loosen the knob, set the cursor for the desired strip thickness, and retighten the knob.

Position your workpiece against the rip fence, and move the fence to bring the left edge of the workpiece against the screw head, as shown in **3–49**. Lock the fence in place, set the jig out of the way, and you're ready to cut a strip, as shown in **3–50**.

After completing the cut, clean up the workpiece on the jointer. Replace the jig in the slot. Then unlock the rip fence, move it to bring the jointed edge against the screw head, lock the rip fence, remove the jig, and saw another strip. Repeat the process as many times as necessary to produce all of the strips that you need for your project.

FOUR-SIDED TAPERING JIG

3–51.

You can taper one side of a table leg without much head-scratching, but tapering all four sides equally presents more of a challenge. With this jig (**3–51**), however, you can cut all four tapers without changing your setup. You simply rotate your workpiece between cuts.

Locate the hold-downs to suit the length of your workpiece. (The pivot block can sit at either end of the jig.) If your tablesaw has a 10" blade, you can handle workpieces up to 2" thick.

Make the Jig

1 Cut a piece of ¾" plywood to the size shown on **3–52**, and then cut a piece of ¼" hardboard to the same dimensions for the base.

2 Cut ⅝" dadoes ³⁄₁₆" deep in one face of the plywood, where dimensioned. Glue the hardboard to the dadoed face with yellow glue. Now, clamp the assembly between two scraps of plywood to ensure even pressure. After the glue dries, remove the clamps, set your dado blade for a ¼"-wide cut, put an auxiliary fence on your miter gauge, and cut a slot through the hardboard centered over each plywood dado, as shown in **3–54**.

3 Cut a piece of maple to ¼ x ⅜ x 12", and then cut two 3" pieces and one 3½" piece from this blank for the guide bars. For the hold-down bases, cut a piece of ¾" plywood to 1½ x 12". Cut a ¼" groove down the center of one face of this plywood, where dimensioned on **3–52**. Drill two ¼" holes near opposite ends of the groove, with each hole centered in the groove and ½" from the end. Cut a 3" piece from each end to make two hold-down bases. Next, glue one guide bar piece in the groove on each hold-down base. After the glue dries, drill a ¼" hole through each assembly, using the previously drilled holes as guides.

4 Cut a maple blank to ¾ x 2 x 12" to make the pivot block. (We begin with an oversized piece to ensure safety during the cutting process.) Cut a rabbet on one end of the blank, where shown on **3–53**. Now, drill two holes to form the ends of the adjustment slot, remove the material between the holes with a coping saw or scrollsaw, and clean up the slot with a file. Cut a ¼" groove centered on the bottom edge of the blank. Next, drill a ¼" hole centered in the groove 2½" from the rabbeted end. Glue in the 3½" guide bar piece, making it flush with the rabbeted end.

After the glue dries, drill a ¼" hole through the blank, using the previously drilled hole as a guide. Trim the blank to 3½" in length. Sand and finish the assembly.

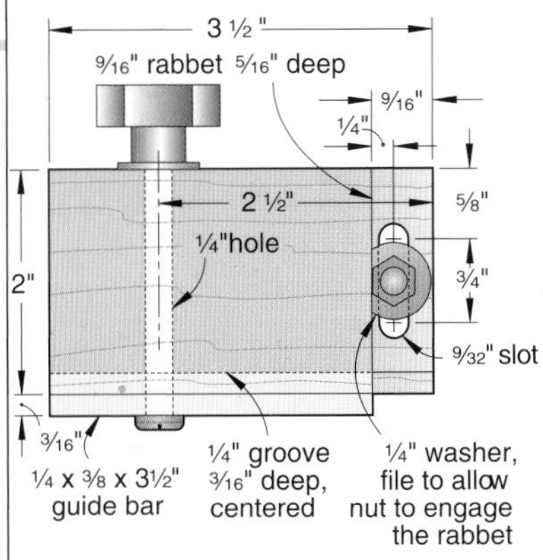
3 ½"
9/16" rabbet 5/16" deep
9/16"
¼"
2 ½"
5/8"
¼"hole
2"
¾"
9/32" slot
3/16"
¼ x 3/8 x 3½" guide bar
¼" groove 3/16" deep, centered
¼" washer, file to allow nut to engage the rabbet
3–53. PIVOT BLOCK

5 Assemble the hold-downs as shown in **3–52**. For the pivot-block, file or grind one edge of the washer flat, as shown on **3–53**, and then assemble the nut, screw, and washer as shown.

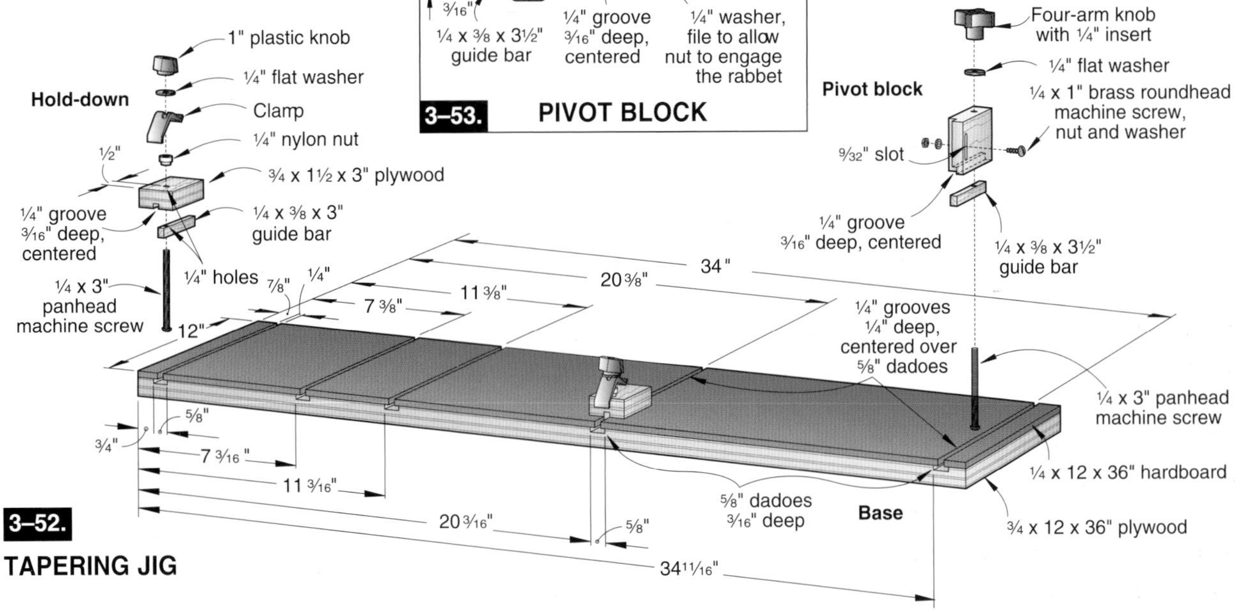
Hold-down
1" plastic knob
¼" flat washer
Clamp
¼" nylon nut
½"
¾ x 1½ x 3" plywood
¼" groove 3/16" deep, centered
¼ x 3/8 x 3" guide bar
¼" holes
¼ x 3" panhead machine screw
12"
7/8"
¼"
7 3/8"
11 3/8"
20 3/8"
34"
Pivot block
Four-arm knob with ¼" insert
¼" flat washer
¼ x 1" brass roundhead machine screw, nut and washer
9/32" slot
¼" groove 3/16" deep, centered
¼ x 3/8 x 3½" guide bar
¼" grooves ¼" deep, centered over 5/8" dadoes
¼ x 3" panhead machine screw
¼ x 12 x 36" hardboard
¾ x 12 x 36" plywood
Base
5/8" dadoes 3/16" deep
5/8"
¾"
7 3/16"
11 3/16"
20 3/16"
5/8"
34 11/16"
3–52.
TAPERING JIG

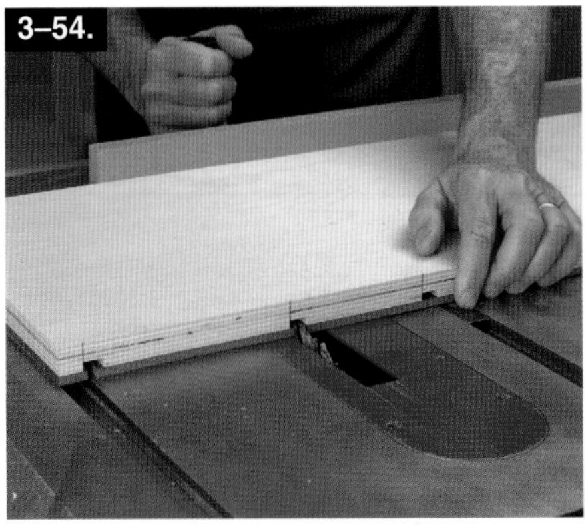
3–54.

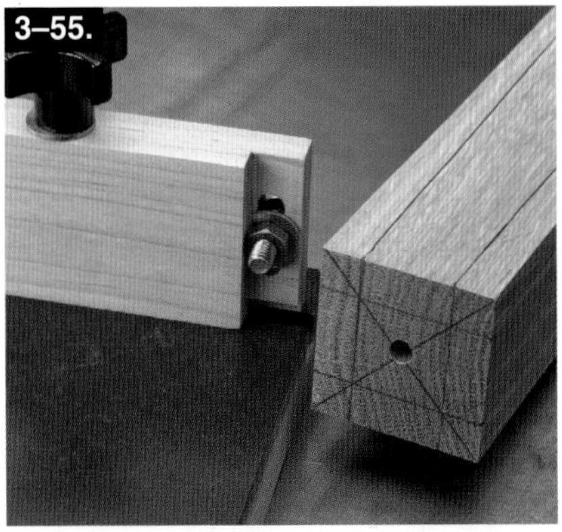
3–55.

Far left: *After cutting dadoes in the plywood base, glue the hardboard to the dadoed face. Mount the two outside blades of a dado set in your tablesaw, and cut slots through the hardboard centered over each dado.*

Near left: *Diagonal lines on the end of the workpiece locate the hole that fits onto the indexing pin. Draw the cutline for the final shape, and extend the lines to the edges to help you position the workpiece on the jig.*

Near right: *Hold the tapering jig tightly against the tablesaw rip fence as you cut. Before starting each pass, make certain that your left hand is well away from the line.*

Far right: *The width and adjustability of the tapering jig allow you to handle a wide range of angle cuts. Here, with the jig flipped end-for-end, we're shaping a simple leg.*

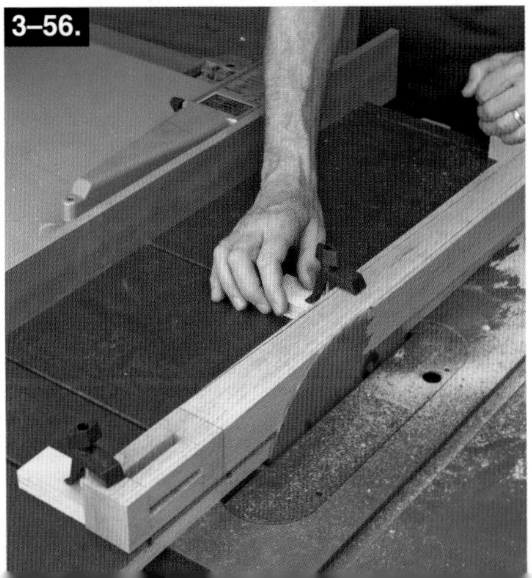
3–56.

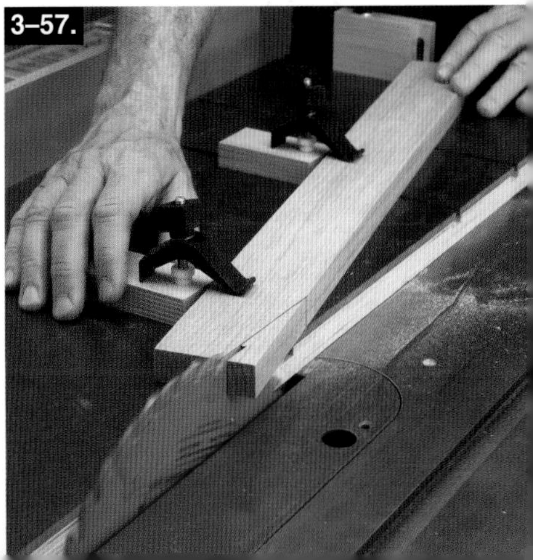
3–57.

Adjustable up or down in the slot, this screw serves as an indexing pin. Once set for a particular workpiece, it guarantees that every cut in the sequence is an equal distance from the center of the workpiece.

Tapering Procedure

To taper a leg, cut your workcpiece to finished length, and then rip it to the square dimensions that you want for the untapered section at the upper end. Draw a line on all four faces to mark where the taper will begin. Drill a ¼" centering hole ⅜" deep at the center of the bottom end, and add cut lines to show the final dimensions of that end, as shown in **3–55**. Draw cut lines on the face connecting the leg-bottom marks with the taper-start marks, as shown in the photo, both to visualize the final shape and to serve as a safety reminder as you push the jig across the saw.

Mount the leg centering hole on the indexing pin. Slide the pivot block until the planned outside face of the leg aligns with the edge of the jig. Turn the knob to lock the pivot block in place. Now, near the upper end of the leg, align the taper-start cutline with the edge of the jig. Slide the hold-down blocks against the leg, and tighten the nylon nut on each one to set the block's position. Tighten the top knob on each hold-down to clamp the leg in place.

Raise the saw blade ¼" above the leg. Butt the jig to the fence, move the fence until the saw blade just clears the left side of the jig, and then make the cut, as shown in **3–56**. To make each of the three remaining cuts, loosen the hold-down knobs, rotate the leg one-quarter turn clockwise (as viewed from the pivoting end), reclamp, and cut.

This jig also serves another purpose, as shown in **3–57**. When you need to cut a single taper, mark its start and stop points on the end and edge of your workpiece. Remove the indexing pin from the end block, and nest the end of the workpiece in the notch. Align the marks with the edge of the jig, and clamp. Place your hold-downs against the workpiece. Tighten the pivot block in place, and make the cut.

90° CROSSCUT SLED

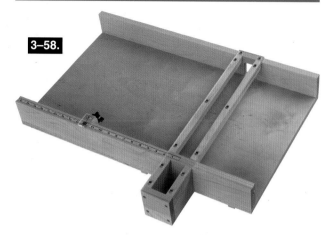

3–58.

A reliable tablesaw miter gauge handles a lot of crosscutting tasks, but not all. It rides in just one slot and supports the workpiece on just one side of the blade, allowing for slop. This problem disappears, however, with a well-made crosscut sled (**3–58**). Making right-angle cutting easier and safer, our design is both simple and econmical to build. And it includes adjustable, reliable stops for repeatable cuts and dead-on accuracy.

Build a Real Workhorse

1 Select a flat piece of ¾" plywood and cut the platform to the dimensions shown on **3–59**.

2 Cut two maple pieces for the fence, and cut a ⅝" groove in the face of one piece, where shown on **3–60**. Glue the two blanks together, keeping the edges flush and the groove on the interior of the lamination. After the glue dries, cut a ¼" groove centered on the ⅝" groove. Then, cut a rabbet along the front of the bottom edge and a ½" groove centered along the top edge.

3 Glue and screw the end to the sides. Now, screw the blade guard to the fence, where shown in **3–59**.

4 Cut the front rail from ¾" maple. Use a jigsaw to cut a notch, where shown in **3–59**, for the blade to pass through. Attach the front rail and the fence to the platform with screws.

3–59.

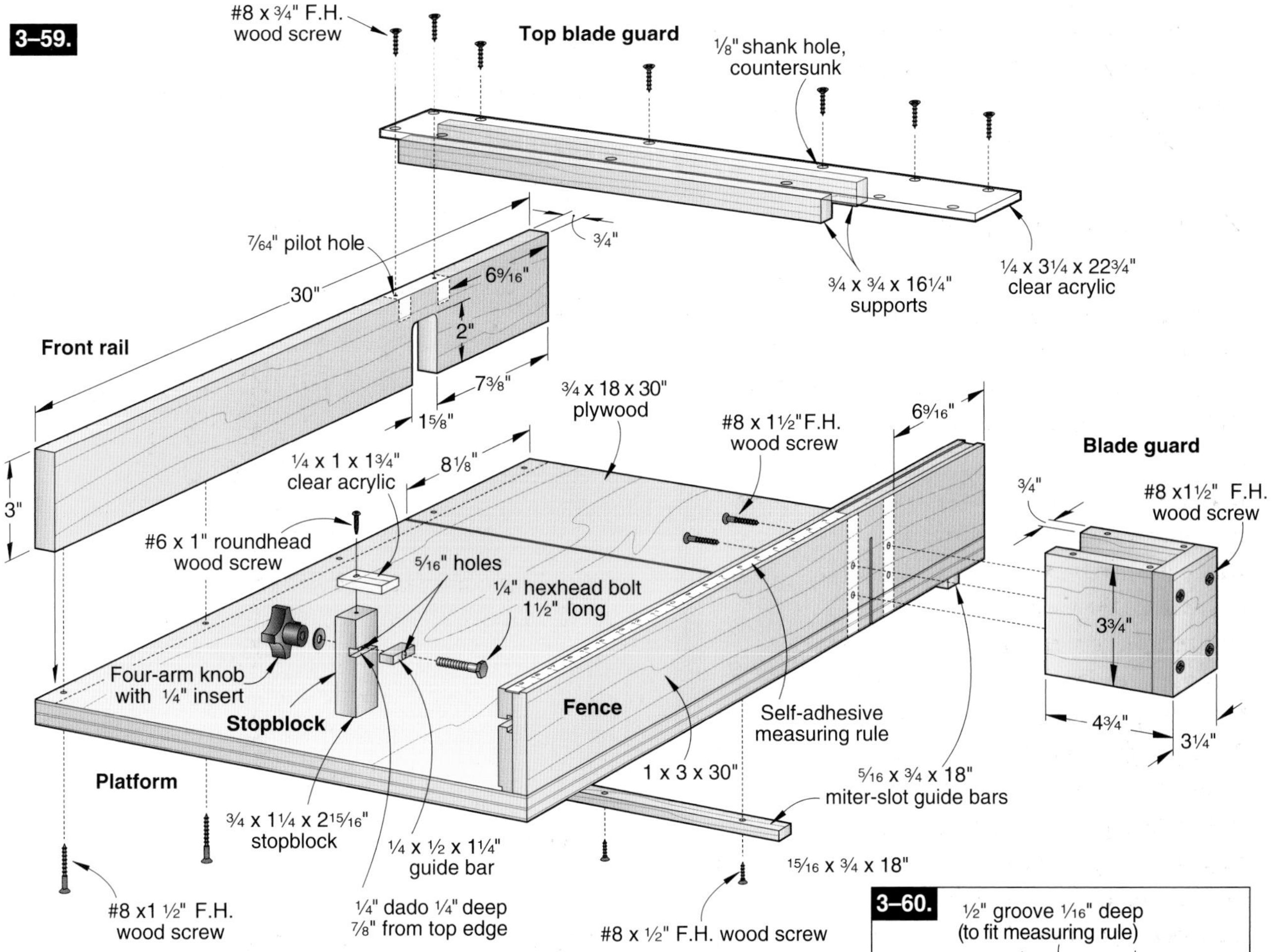
#8 x 3/4" F.H. wood screw
Top blade guard
1/8" shank hole, countersunk
7/64" pilot hole
3/4"
6 9/16"
30"
2"
Front rail
7 3/8"
1 5/8"
3/4 x 3/4 x 16 1/4" supports
1/4 x 3 1/4 x 22 3/4" clear acrylic
3/4 x 18 x 30" plywood
#8 x 1 1/2" F.H. wood screw
6 9/16"
Blade guard
1/4 x 1 x 1 3/4" clear acrylic
8 1/8"
3/4"
#8 x1 1/2" F.H. wood screw
3"
#6 x 1" roundhead wood screw
5/16" holes
1/4" hexhead bolt 1 1/2" long
3 3/4"
Four-arm knob with 1/4" insert
Stopblock
Fence
Self-adhesive measuring rule
4 3/4"
3 1/4"
Platform
1 x 3 x 30"
5/16 x 3/4 x 18" miter-slot guide bars
3/4 x 1 1/4 x 2 15/16" stopblock
1/4 x 1/2 x 1 1/4" guide bar
15/16 x 3/4 x 18"
#8 x1 1/2" F.H. wood screw
1/4" dado 1/4" deep 7/8" from top edge
#8 x 1/2" F.H. wood screw

5 Cut, sand, and finish two top blade guard supports. Using a fine-toothed tablesaw blade, cut a piece of 1/4" clear acrylic to size for the guard cover. Attach the cover to the supports, the front rail, and the fence.

6 From 3/4" maple stock, cut two strips to serve as miter-slot guide bars. Set your tablesaw rip fence 8 1/8" to the right of the blade, and lower the blade below the table's surface.

*(**Note:** Make sure your fence is parallel to the miter gauge slot before proceeding.)*

Apply double-faced tape to the top of each guide bar, and attach the bars to the platform, as shown in **3–62** and **3–63**. Remove the assembly from the saw, and permanently attach the bars with screws.

7 Cut a piece for the stopblock, and cut a dado in the back, where shown. Cut a guide bar, and glue it into the dado. Drill a shank hole through the block and bar, where shown. Now, cut a piece of 1/4" acrylic plastic to size for the stopblock indicator (**3–61**). Drill, saw, and file smooth the slot, where shown. Make a cursor line, as shown in **3–48**, on *page 85*.

3–60.

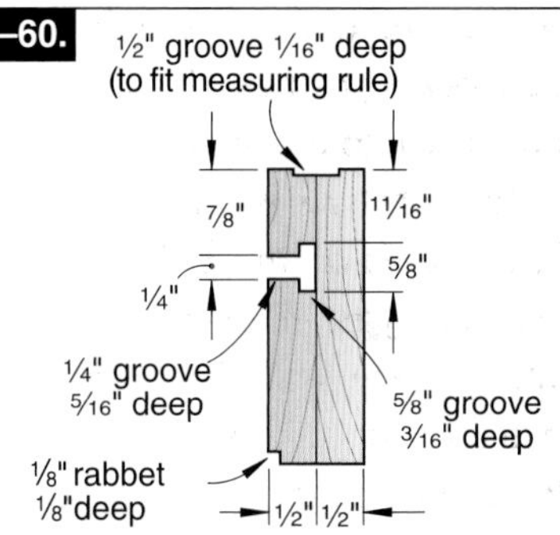
1/2" groove 1/16" deep (to fit measuring rule)
7/8"
11/16"
5/8"
1/4"
1/4" groove 5/16" deep
5/8" groove 3/16" deep
1/8" rabbet 1/8" deep
1/2"
1/2"

FENCE SECTION VIEW

3–61.

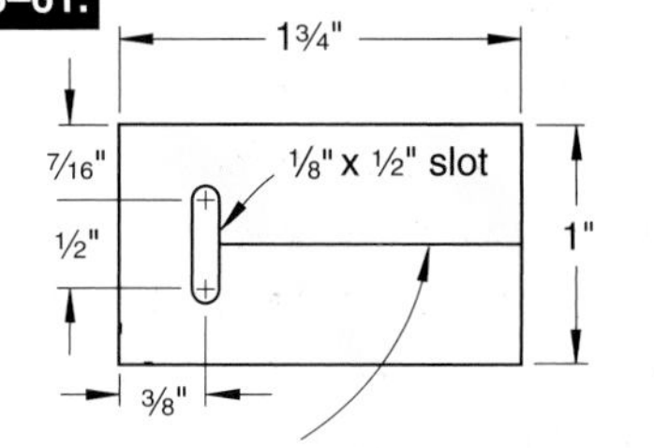
1 3/4"
7/16"
1/8" x 1/2" slot
1/2"
1"
3/8"

Score a line on the acrylic with a knife, and color it with a permanent marker.

INDICATOR DETAIL

3-62.

Two pennies shim the miter-slot guide bars slightly above the tablesaw surface. Place a couple of these stacks in each miter-gauge slot, and set the bars on top.

3-63.

Keeping the right end of the platform against the rip fence, set the sled assembly on the guides. Press down firmly to stick the bars to the platform.

3-64.

Hold the workpiece firmly against the fence as you make a cut. Keep your hands outside the blade guard, and don't cut through its end.

8 Remove the top blade guard, sand the jig, and apply three coats of finish. Reattach the blade guard, assemble and install the stopblock, place the crosscut sled on your tablesaw, and make a cut from the front edge through the fence. Use a rule to set the stopblock 4" from the kerf. Mark the center of the stop block on its top end, align the 4" line on the self-adhesive measuring tape with that mark, and attach the tape in the fence groove. Use tin snips to cut off the portion of the tape extending beyond the left end of the fence. Place the indicator on the stopblock, align the cursor with the tape's 4" line, and attach the indicator to the block with a screw.

Now, Go Sledding

If a workpiece fits between the fence and the front rail, you can cut it on your crosscut sled, as shown in **3-64**. Use the stop block to cut multiple pieces to the same length, provided that length falls within the stop block's range. Remove the stopblock when cutting pieces that extend beyond that range. When you install a tablesaw blade of a different thickness or with a different tooth set than the one used to calibrate your stopblock, check the setting with a rule and adjust the cursor.

3-65.

The key to this super-simple box-joint jig is setting it up precisely so the "fingers" of your box joints line up regardless if the workpiece is 3" or 13" wide.

Basic box-joint jigs like this one (**3-65**) have been around for a long time. What's different here is that we tell you the secrets for setting the jig up for perfect results! After you finish building and adjusting the jig, see *page 93* for tips on using it.

Note: *To use this jig you will need a tablesaw, stackable dado set, and a calipers with dial or digital readout. You can purchase the dial type for $15 to $40; digital models cost $60 to $75.*

We've found dial calipers essential because box joints must be cut with exacting accuracy. Why? Any minor error in the width or spacing of the individual "fingers," even .001", multiplies itself with every finger you cut.

For reasons of design and proportion, you typically make the individual "fingers" in box joints as wide as the thickness of the

3–66.

workpieces. For this, we'll make a jig for cutting ¼"-wide fingers in ¼"-thick stock. (See **3–66** for the precise dimensions of the joints.) For thicker or thinner stock, you adjust the size of the jig's pin and the width and height of the dado cut accordingly.

Make the Jig

1 Cut parts A, B, and C to the sizes shown in **3–67**. Any flat and straight stock will do.

2 Adjust your dado set for a cut that's .001" wider than the joint fingers (.251" in our example). With stackable dado sets,

you can place commercially made shims between the cutters, or make your own shims from various papers. (Standard tablet paper measures .002 to .005" thick, some tissue and waxed papers measure .001" thick.) Check your adjustment by measuring a test cut with your calipers, as shown in **3–68**. Raise the dado set ½" above the tabletop.

3 Cut the notch that holds the pin in part A. Do this by holding part A against the miter gauge with part B beneath it, as shown in **3–69**. Do not cut into part B.

4 Cut a ¼ × ¼ × 6" strip of hardwood that fits snugly into the notch you just cut in part A. (The strip should slip into place, yet fit tightly enough so it doesn't fall out.)

3–68.

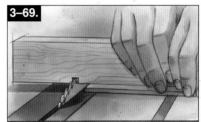

3–69.

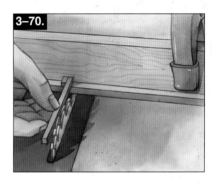

3–70.

Cut a 1½" pin from the strip and glue it into the notch, flush with the back of part A. Save the leftover strip. Screw part B to part A.

5 Set your miter gauge for a 90° cut. Use the leftover strip to position the jig assembly on the miter gauge. Do this by aligning the jig pin ¼" from the path of the dado set, as shown in **3–70**. With the pin aligned, temporarily clamp the jig to the miter gauge, and then affix the gauge to the jig with screws. Replace the miter

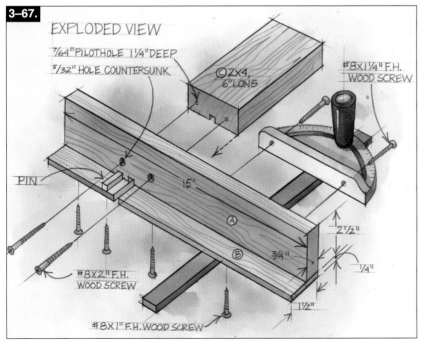

3–67.

EXPLODED VIEW
7/64" PILOT HOLE 1¼" DEEP
5/32" HOLE COUNTERSUNK
©2X4, 6" LONG
#8X1¼" F.H. WOOD SCREW
PIN
15"
Ⓐ
Ⓑ
2½"
¾"
¼"
1½"
#8X2" F.H. WOOD SCREW
#8X1" F.H. WOOD SCREW

3–71.

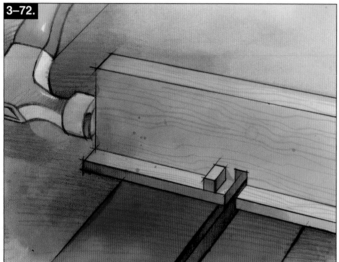

3–72.

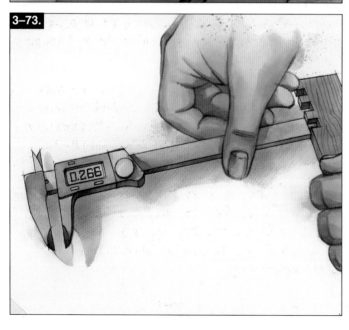

3–73.

0.266

gauge into its slot and cut through parts A and B. Attach the blade guard (C) centered behind the notch you just cut.

During this and the following steps, apply pressure to the miter gauge to hold its bar firmly against the right side of the slot. This will keep its distance from the dado set consistent during cuts.

6 Position a piece of scrap stock as shown in **3–71**, and cut a notch into the scrap piece. Position this notch over the pin and make another cut. Position that notch over the pin and repeat the cut.

With the calipers, check the width of the fingers. They should be .001" under your desired finger width. (For our ¼" fingers, the calipers should read .249".)

7 Chances are your jig will need some adjustment to achieve the necessary finger width. If the fingers are too wide, say .255" in our example, tap the end of the jig closest to the blade with a hammer, as shown in **3–72**. Make more test cuts and tapping adjustments as necessary. If the fingers are too narrow (.245" in our example), tap the other end of the jig. Even though the jig is screwed in place, the hammer taps will make these fine adjustments.

8 With your calipers, check the depth of the fingers in your scrap stock (3–73). Adjust the height of your blade until the depth reads .016" more than the width of your fingers (.266" in our example). This leaves the fingers long enough so you can sand them flush with the box later.

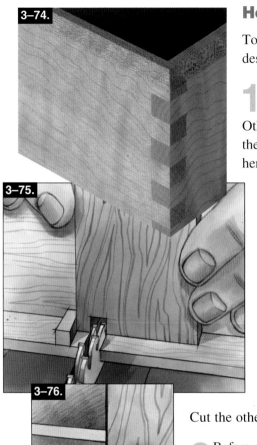

3–74.

3–75.

3–76.

3–77.

SCRAP

SCRAP

SCRAP

How to Make Snug, Good-Looking Box Joints

To master the art of making box joints (**3–74**) using the jig just described, do the following:

1 Before you cut the actual box joints, keep in mind that the width of your box sides must be an increment of the finger width. Otherwise, you'll wind up with less-than-pleasing partial fingers at the bottom of the box. So, in the example of ¼"-thick stock discussed here and in the information on building the jig, the width of the workpieces must be an increment of ¼" (such as 5", 5¼", 5½", etc.).

2 Mark the front, back, and side pieces of your box. Also, mark the top edge of each of these pieces.

For each box you make, you cut the sides, the front, and the back consecutively. It doesn't matter which pair you do first, so we'll start with the sides.

For all of the following cuts, hold the top edge of the workpiece toward the jig pin for the first cut. Now, put hand pressure on the jig to hold its miter-gauge bar firmly against the right side of the tablesaw channel. Make the first cut, as shown in **3–75**. Place the just-cut notch over the jig pin and repeat to cut fingers along the full width of the workpiece. Cut the other side piece in the same fashion.

3 Before you cut the front and back, cut one notch into a scrap piece, just as you cut the first notch into the side piece earlier. Position this notch over the pin as shown in **3–76**, and butt the top edge of the front piece against the scrap before making a cut.

Make the remaining cuts in the front piece by removing the scrap, placing the notch over the pin, and proceeding as described earlier. Cut the back piece as you cut the front.

4 To join your box pieces, apply glue to all of the mating surfaces with a small brush. Tap the joints together with a rubber mallet if necessary. Clamp the box together, as shown in **3–77**. (You may need to position a clamp diagonally to square the box.)

Note that we used scrap pieces on each side of the corners to evenly distribute the clamping pressure along the joint. Wider boxes may require additional clamps.

After the glue dries, sand the fingers flush with the sides, front, and back. Be careful not to round over the corners.

3–78.

3–79.

3–81.

MITER JIGS

Cutting accurate 45° miters becomes a whole lot easier with these angled tablesaw-miter jigs (**3–78** and **3–79**). The key, of course, is to lay out the fences with precision while making the jigs. The jig marked R on **3–80** fits on the right-hand side of the

blade, as shown in **3–78**. The jig marked L, shown in **3–79**, fits into the miter-gauge slot on the left-hand side of the tablesaw blade. Use either jig for miter-cutting 45° angles. Or, turn the jig marked L end for end, and use that fence when crosscutting at 90°, as shown in **3–79**.

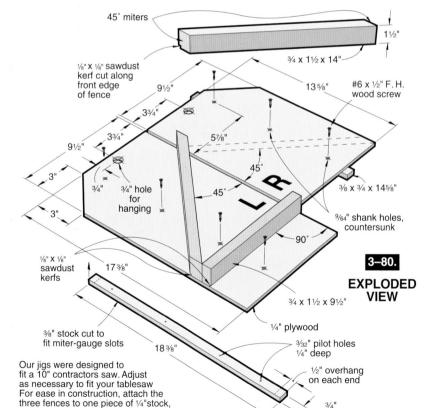

45° miters

⅛" X ⅛" sawdust kerf cut along front edge of fence

9½"

3¾"

3¾"

9½"

3"

¾"

3"

¾" hole for hanging

5⅞"

45°

45°

90°

L R

13⅝"

1½"

¾ x 1½ x 14"

#6 x ½" F. H. wood screw

⅜ x ¾ x 14⅝"

⁹⁄₆₄" shank holes, countersunk

3–80.
EXPLODED VIEW

⅛" X ⅛" sawdust kerfs

17⅜"

18⅜"

¾ x 1½ x 9½"

¼" plywood

3/32" pilot holes ¼" deep

½" overhang on each end

¾"

⅜" stock cut to fit miter-gauge slots

Our jigs were designed to fit a 10" contractors saw. Adjust as necessary to fit your tablesaw For ease in construction, attach the three fences to one piece of ¼"stock, and then cut the ¼" stock in half where shown on the drawing.

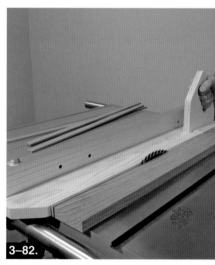

3–82.

Cutting wedges and ripping thin strips rank among the more dangerous tasks you can perform on a tablesaw. Our two ripping jigs (**3–81** and **3–82**) add an extra measure of safety to these operations, plus you only position the fence once to make multiple pieces.

You can build these jigs from any flat pieces of scrap wood.

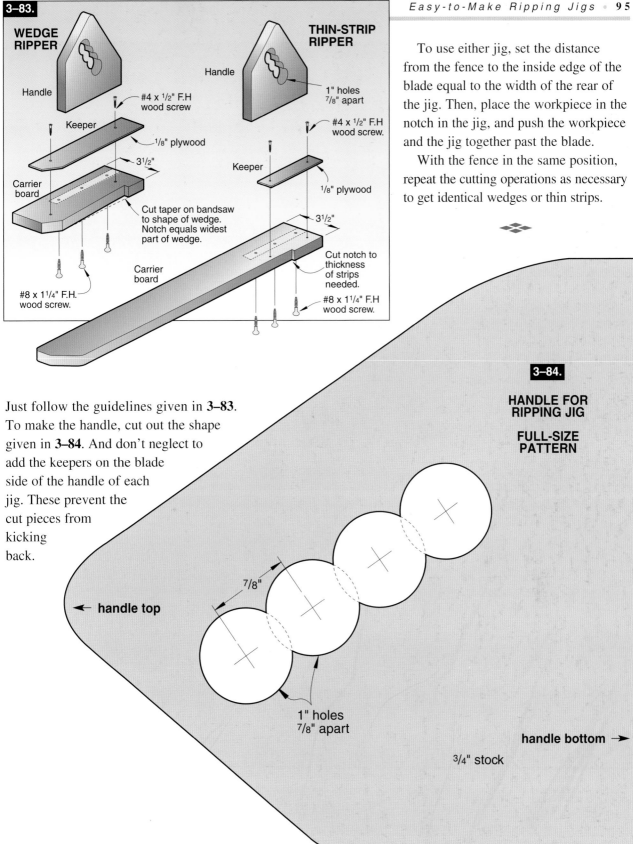

3–83.

WEDGE RIPPER

Handle

Handle

#4 x ¹/₂" F.H wood screw

Keeper

¹/₈" plywood

3¹/₂"

Carrier board

Cut taper on bandsaw to shape of wedge. Notch equals widest part of wedge.

#8 x 1¹/₄" F.H. wood screw.

Carrier board

THIN-STRIP RIPPER

Handle

1" holes ⁷/₈" apart

#4 x ¹/₂" F.H wood screw.

Keeper

¹/₈" plywood

3¹/₂"

Cut notch to thickness of strips needed.

#8 x 1¹/₄" F.H wood screw.

To use either jig, set the distance from the fence to the inside edge of the blade equal to the width of the rear of the jig. Then, place the workpiece in the notch in the jig, and push the workpiece and the jig together past the blade.

With the fence in the same position, repeat the cutting operations as necessary to get identical wedges or thin strips.

Just follow the guidelines given in **3–83**. To make the handle, cut out the shape given in **3–84**. And don't neglect to add the keepers on the blade side of the handle of each jig. These prevent the cut pieces from kicking back.

3–84.

HANDLE FOR RIPPING JIG

FULL-SIZE PATTERN

handle top

⁷/₈"

1" holes ⁷/₈" apart

handle bottom

³/₄" stock

THIN-STRIP RIPPING PUSHBLOCK

Safely rip thin strips between your tablesaw blade and rip fence with this fence-supported pushblock (**3–85** and **3–86**). The replaceable end block allows you to push the thin piece being cut completely through the cutting area, eliminating the chance of kickback.

Use a hand plane handle for a full-size pattern to mark the outline for the pushblock handle. We've located the handle up and out of the way, so your fingers remain safely away from the blade during the cutting operation.

3–85.

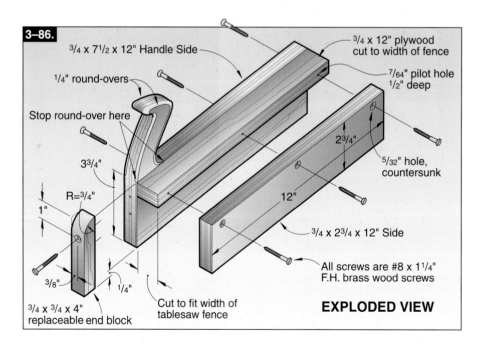

3–86.

3/4 x 7 1/2 x 12" Handle Side

1/4" round-overs

Stop round-over here

3 3/4"

R=3/4"

1"

3/8"

1/4"

3/4 x 3/4 x 4" replaceable end block

Cut to fit width of tablesaw fence

3/4 x 12" plywood cut to width of fence

7/64" pilot hole 1/2" deep

2 3/4"

5/32" hole, countersunk

12"

3/4 x 2 3/4 x 12" Side

All screws are #8 x 1 1/4" F.H. brass wood screws

EXPLODED VIEW

3–87.

PRECISION MITER STOP

Use this handy stop (**3–87** and **3–89**) on your own 2⅝"-wide miter-gauge extension, or add it to the sliding tablesaw jig shown in **3–88**. It sits on the fence and allows you to cut piece after piece to the same length.

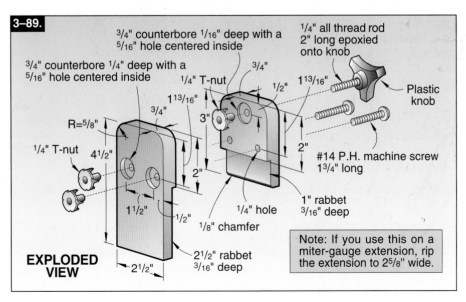
3–88.

3–89.

¾" counterbore 1/16" deep with a 5/16" hole centered inside

¾" counterbore ¼" deep with a 5/16" hole centered inside

¼" all thread rod 2" long epoxied onto knob

¼" T-nut

Plastic knob

1 13/16"

¾"

½"

3"

R=⅝"

4½"

¼" T-nut

1 13/16"

2"

2"

#14 P.H. machine screw 1¾" long

1½"

½"

¼" hole

1/8" chamfer

1" rabbet 3/16" deep

2½" rabbet 3/16" deep

2½"

EXPLODED VIEW

Note: If you use this on a miter-gauge extension, rip the extension to 2⅝" wide.

CHAPTER 4

Bandsaw Aids

IF YOU OWN ONE, YOU KNOW THAT *a bandsaw really comes in handy. With it, you can quickly cut curves and circles (and the smaller the blade, the tighter the circle), make angled cuts by tilting the table, and resaw thick stock into thinner stock. The jigs, fixtures, and accessories in this chapter will help you more efficiently set up your saw to do all those things, and do them with greater accuracy. You also may discover some jigs that allow you to do even more cutting operations. Begin with the easy-to-make bandsaw fence that follows.*

PRECISION FENCE

Bandsaws are like after-dinner speakers: They're a lot more enjoyable if they don't wander too far off course. Here's the jig you need to produce nice, straight cuts (**4–1**). (And it mates perfectly with the multi-jig on *page 103*.) Begin by cutting the pieces and assembling them as shown in **4–2**. Once you have all the parts assembled, position the fence on your bandsaw and tighten the ⅜" handle to secure the fence in place. Test-rip on a piece of

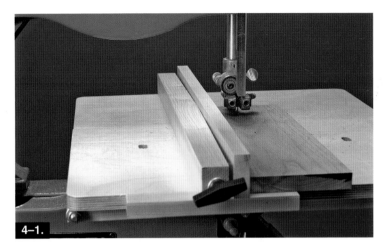

4–1.

scrap, and alternately loosen one machine screw and tighten the other until the fence is parallel to the cutting track of the blade.

Be sure to loosen and tighten the screws the same amount to avoid bowing the fence.

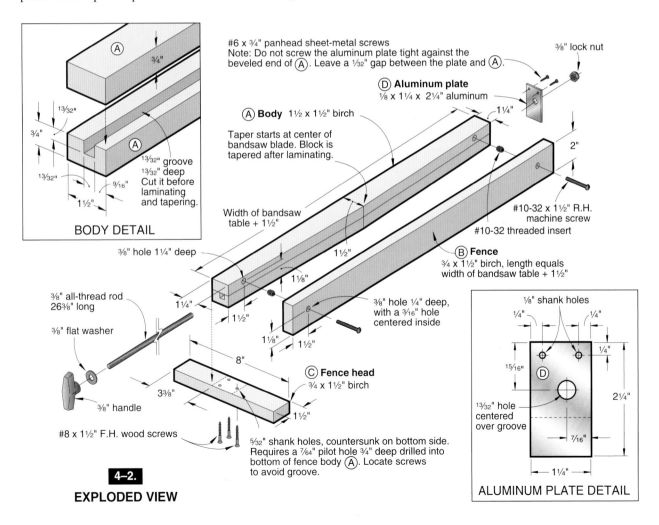

BODY DETAIL

(A) — 3/4"

13/32" — 3/4" — 13/32"

(A) — 13/32" groove 13/32" deep Cut it before laminating and tapering.

9/16" — 1½"

#6 x ¾" panhead sheet-metal screws
Note: Do not screw the aluminum plate tight against the beveled end of (A). Leave a 1/32" gap between the plate and (A).

⅜" lock nut

(D) **Aluminum plate**
⅛ x 1¼ x 2¼" aluminum

1¼"

(A) **Body** 1½ x 1½" birch

Taper starts at center of bandsaw blade. Block is tapered after laminating.

Width of bandsaw table + 1½"

2"

#10-32 x 1½" R.H. machine screw
#10-32 threaded insert

(B) **Fence**
¾ x 1½" birch, length equals width of bandsaw table + 1½"

⅜" hole 1¼" deep

⅜" all-thread rod 26⅜" long

⅜" flat washer

1¼"

1½"

1⅛"

1⅛" 1½"

⅜" handle

8"

3⅜"

1½"

#8 x 1½" F.H. wood screws

(C) **Fence head**
¾ x 1½" birch

⅜" hole ¼" deep, with a 3/16" hole centered inside

5/32" shank holes, countersunk on bottom side. Requires a 7/64" pilot hole ¾" deep drilled into bottom of fence body (A). Locate screws to avoid groove.

4–2.

EXPLODED VIEW

ALUMINUM PLATE DETAIL

⅛" shank holes

¼" ¼"

¼"

15/16" (D)

13/32" hole centered over groove

2¼"

7/16"

1¼"

RESAWING JIGS

Your lumber may measure $^{13}/_{16}$" or $^3/_4$" thick when you bring it home, but it doesn't have to stay that way. Maybe you need $^1/_2$" pieces for drawer sides, or you want to slice a beautifully figured board into veneer for a jewelry box. No problem. With your bandsaw, you can resaw a board quickly to any thickness that you want; and this pair of jigs (**4–3**) makes it a whole lot easier. (For an even simpler resawing guide, though not for making veneer, see *page 102*.)

For trouble-free resawing on the bandsaw, start with the right equipment. Bandsaw expert Mark Duginske recommends a $^1/_2$"-wide hook-tooth blade with three teeth per inch for saws with more than $^1/_2$ hp. Less-powerful machines benefit from a blade with four teeth per inch.

If your blade is dull, buy a new one. A sharp blade goes a long way toward eliminating drift, which is the tendency of a blade to cut at a slight angle, rather than parallel to the miter-gauge slot. Then, test the blade tension.

Make the Jigs

The unique shape of the fence for this resawing jig allows you to keep the bandsaw's blade guides as close together as possible during every cut. That support minimizes any twisting and flexing in the blade, resulting in the truest cut

4–3.

When you resaw to make thin veneer, this jig's high, flat fence puts your mind at ease. Add the feather board, and you have all of the support you need.

4–4.

RESAWING JIG

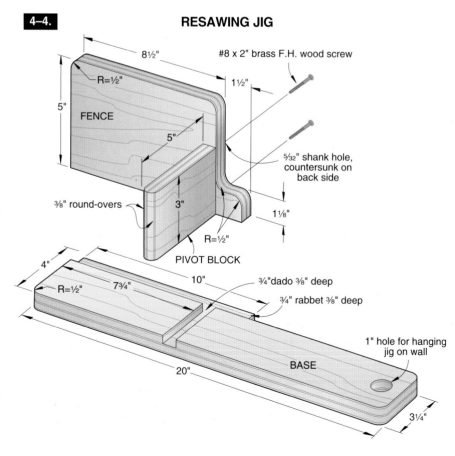

4–5.

FEATHER BOARD

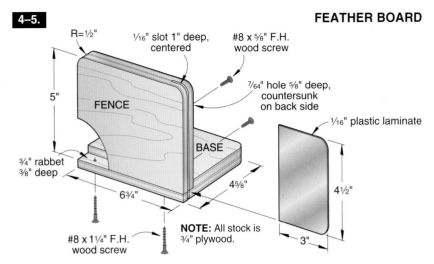

R=½"

¹⁄₁₆" slot 1" deep, centered

#8 x ⅝" F.H. wood screw

⁷⁄₆₄" hole ⅝" deep, countersunk on back side

5"

FENCE

¹⁄₁₆" plastic laminate

BASE

¾" rabbet ⅜" deep

6¾"

4⅝"

4½"

#8 x 1¼" F.H. wood screw

NOTE: All stock is ¾" plywood.

3"

possible. This two-piece jig, shown in **4–4** and **4–5**, is built from ¾" birch plywood. To make it, follow these six steps:

1 Make the base of the long unit by cutting a 4 × 20" piece of plywood. Then cut a ¾" rabbet ⅜" deep along the top edge, where shown in **4–4**.

2 Cut a dado 7¾" from the left end. Measure 10" from the left end, and mark the rabbet. Go to the bandsaw, and trim off the portion of the rabbet to the right of the mark. Drill a 1" hole at the right end of the base, just to make the jig easy to hang on the wall when it's not in use. Cut the pivot block to size. Rout ⅜" round-overs on one edge, where shown in **4–4**.

3 For the fence, begin by cutting a piece 10" long and narrow enough to fit between your bandsaw table and the upper wheel housing, while resting on the base rabbet. (We made ours 5" wide.)

4 At the bandsaw, shape the leading end of the fence, as shown in **4–4**. Cut ½" radii on the corners of the fence and base, where shown.

5 Glue and clamp the three pieces together, and allow the glue to dry. For added strength, drive two brass screws through the fence and into the pivot block, as shown in **4–4**.

6 Finally, make the simple feather board from plywood and plastic laminate. Cut the base to the dimensions shown in **4–5**, and rabbet one edge. Cut the fence, and then put a 1"-deep kerf in one end with the bandsaw. Cut a ½" radius on each of the top corners. Cut the plastic laminate, slip it into the kerf, and fasten it with two ¾" screws. Attach the fence to the base with glue and a couple of screws driven

You can use the jig's pivot block for most resawing jobs except making veneer. The rounded end makes it easy to adjust for blade drift.

from below. Brush three coats of polyurethane on both jigs to protect the wood.

Set Up and Resaw

1 Check that your bandsaw table is set at 90° to the blade, and then put the jigs to work.

2 You need adequate tension on your bandsaw blade to get top results. To adjust the tension, set the upper guide 6" above the table. Push on the side of the blade with your little finger about 3" above the table. If the blade deflects more than ¼" under moderate pressure, increase the tension.

3 To resaw a board in half or simply cut a slab from a thicker piece, set up the pivot block on your resaw jig. Mark a guideline on the top edge of your stock with a pencil. Use that line to position your resaw jig. Clamp the jig on the left side of the blade, as shown in **4–6**. Use an adjustable bar clamp at each end of the base, and locate the pivot

4–6.

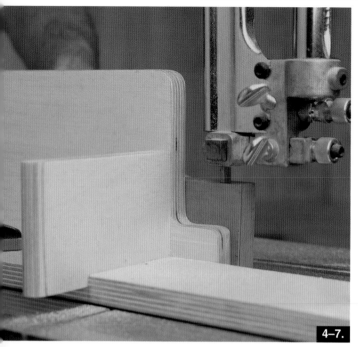

4-7.

The cut-out section of the resawing fence allows you to bring the upper blade guides down as low as possible. This step keeps the blade from twisting as you resaw.

block alongside the blade, leaving a gap equal to the distance between your stock's left face and the cutline.

4 Set the upper guide of the bandsaw ¼" above the upper edge of your stock. Hold the left side of your workpiece tightly against the pivot block, and slowly feed it into the blade. The rounded end of the block allows you to slightly adjust the angle of the board as you saw, compensating for any blade drift.

5 Use a scrap wood pushstick as you complete the cut. That step keeps your hands safely away from the blade.

Let's Make Some Veneer

When you resaw very thin layers of wood, it helps to have plenty of support on both sides of your stock. You can get that support by using the fence of the resawing jig and the feather board as follows:

1 Position the fence at a distance from the blade equal to the desired thickness of your veneer, and

extending about ½" past the back edge of the blade. Clamp the jig to the bandsaw table, front and back.

2 Angle the feather board, with the laminate wing on the near side of the blade and slightly closer to the fence than the width of your stock. Secure it to the table with two clamps.

3 Lower the upper guide of the saw to within ¼" of the workpiece, and resaw slowly, allowing the blade time to cut properly. If you're cutting several pieces from the same stock, gradually making it thinner, make sure that the feather board continues to hold the workpiece snugly against the fence. If drift is a problem, adjust the angle of the fence to suit.

The narrower the workpiece, the more value you get from the unique design of the jig. The cut-out portion allows you to lower the upper guide to suit stock of any width, as shown in **4–7**, while retaining the support of the broader fence side.

4 Even the best resaw job creates some roughness on the veneer and the original stock. So, before sawing another piece of veneer from the stock, run the remaining material across the jointer. This step ensures that you're again working with a smooth surface against the fence.

FULLY ADJUSTABLE RESAWING GUIDE

This resawing guide (**4–8** and **4–9**) lets you correct for blade drift, and you can build it from parts you probably have lying around your shop.

After struggling with his bandsaw fence, blocks, clamps, and a resaw guide, *WOOD*® magazine reader John Hodges of Kaufman, Texas, decided to design his own bandsaw resawing guide. You can build one just like it by gathering up some scrap stock and following **4–9**.

To use this guide, first mark a line along the top edge of the piece to be resawn. Adjust the center

4–8.

to B. Tighten the wing nut in part C to secure it in the miter-gauge slot.

Because few bandsaw blades track perfectly straight (making a fence almost useless for resawing), the curved end of part A allows you to steer the board into the bandsaw blade and make adjustments to follow your marked line. We recommend using a ½"- to ¾"-wide skip-tooth or hook-tooth blade for cleaner cuts. And, always use a pushstick for safety when resawing on a bandsaw.

portion of the jig (A) until the bandsaw blade aligns with the marked line on the wood. Tighten the wing nuts that hold A securely

MULTI-JIG THAT STOPS BLADE WANDER

If your bandsaw isn't the precision multi-faceted tool you want it to be, we've got just the solution for you (**4–10**). The jig table increases the size of your bandsaw table, and the guides steady the blade at tabletop height to minimize blade wander and increase accuracy. The auxiliary circle-cutting guide attachment allows you to cut disc after disc with amazing accuracy and consistency without having to drill a hole in your workpiece. And finally, add the easy-to-align fence shown on *page 99* for precision ripping.

Note: This jig was designed to fit most 14" bandsaw tables. If your metal bandsaw table measures more than 10" in front of or behind the blade, you'll need to increase the depth (front to back) of the jig. If your metal table measures more than 9" on either side of the blade, you'll need to increase the width of the jig table. For larger bandsaw tables, the distance between the guides (E) must be ¼" more than the width of your metal table plus the length of the protrusion of the metal alignment pin.

4–9.
RESAWING GUIDE

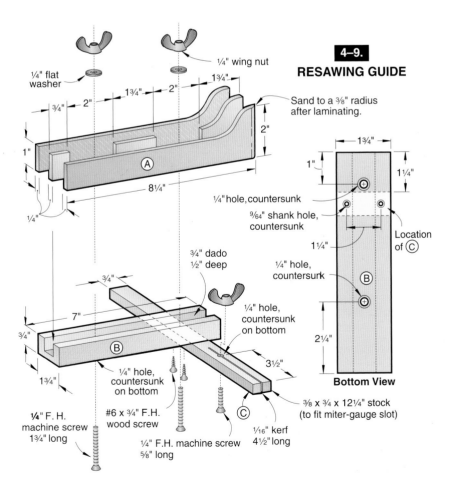

4-10.

webs or brackets on the bottom side of the metal table. Hold a drill with a ¼" high-speed twist bit in it directly over the marked centerpoints, and make sure the drill can be positioned so the bit is perpendicular to the table. If the top of the bandsaw gets in the way, move the hole centerpoints forward, again being careful not to locate them over any webs or brackets on the bottom side of the metal table.

4-11.
SPACER/GUIDE
(right-hand spacer/guide shown)

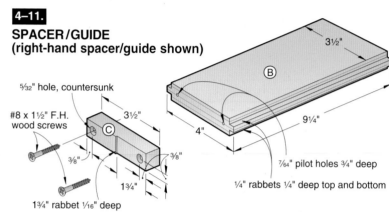

Start with the Two-Part Plywood Table

1 Cut the two tables (A) to the size listed in the Materials List. (Due to its stability and strength, we used ¾" [18mm actual] Baltic birch plywood.) Mark and cut a ¾" radius on two corners of each table, where shown on **4–15**, on *page 106*.

2 Using a ¼" spline cutter in a router table or a dado blade in your tablesaw, cut a ¼" groove ⁵⁄₁₆" deep, centered along the inside edge of each table (A).

3 Place the front table (A) on your metal bandsaw table, and center the table side-to-side from the blade. Now, stand in front of your bandsaw, and move the table toward you until it is 1¾" in front of the center of the bandsaw blade. Clamp the table in place.

4 Look under the metal table on your bandsaw, and locate any protruding webs or brackets. Then, locate and mark the centerpoints for a pair of ¼" holes on the front table, being careful not to locate the holes directly over any

4-12.
EXPLODED VIEW
(left-hand spacer/guide shown)

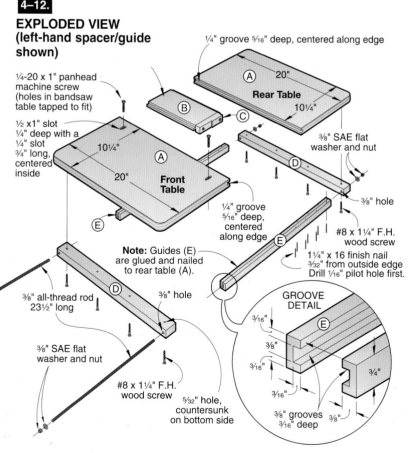

Materials List for Multi-Jig

PART	FINISHED SIZE T	W	L	MTL.	QTY.
A tables	¾"	10¼"	20"	BP	2
B spacer guides	¾"	4"	9¼"	BP	2
C blade guides	¾"	¾"	3½"	B	2
D clamp blocks	¾"	1¼"	20"	B	2
E guides	¾"	¾"	19⅞"	B	2

Materials Key: BP = birch plywood;
B = birch.
Supplies: Eight #8 x 1¼" flathead wood
screws; four #8 x 1½" flathead wood crews;
1¼" x 16 finish nails; two ⅜" all-thread rod,
23 ½" long; four ⅜" SAE flat washers and
nuts; two ¼–20 x 1" panhead machine
screws.

5 Remove the front table
section from your bandsaw jig,
and form a pair of counterbored
slots in it, where indicated on the
Front Table drawing in **4–15**. The
slots allow you to adjust the jig
table on the metal bandsaw table.

6 Reposition and clamp the
front table (A) to your bandsaw table 1¾" in front of the
center of the blade. As shown in
4–13, use a centerpunch to mark
the *center* of the slot locations
onto the metal table. Remove the
front wood table, and drill a ¹³⁄₆₄"
hole through your metal bandsaw
table at each marked centerpoint.

7 Tap the holes in the metal
bandsaw table with a ¼–20 tap.

Next, Add the Spacers and Blade Guides

1 Cut the spacers (B) and blade
guides (C) to size.

2 Using your tablesaw or router
table, form a tongue along
both edges of each spacer to fit
snug but slide smoothly in the
mating grooves in the front and
rear tables (A).

3 Form a ¹⁄₁₆"-notch 1¾" long
in each blade guide, where
shown on **4–11**.

4 Drill the holes, and screw
the blade guides (C) to the
spacers (B). Note that you will
have a right-hand and left-hand
spacer/guide. Sand the top and
bottom of the blade guides flush
with the spacers if necessary.

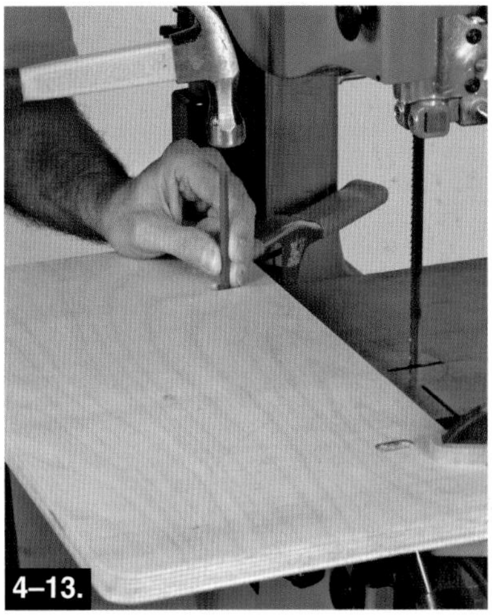

4–13.

*Using the front table as a guide,
use a centerpunch to mark the hole
locations on the metal bandsaw table.*

Cut and Secure the Clamp Blocks and Guides

1 Cut the clamp blocks (D) to
size. Mark and drill the ⁵⁄₃₂"
holes, where shown on **4–15**.

2 Drill a pair of ⅜" holes in
each clamp block for the
all-thread rod to pass through.

3 Cut two pieces of ⅜" all-
thread rod to 23½" long.

4 Screw the blocks to the *bottom
side* of the front and rear
tables (A), where shown on **4–15**.

5 To form the guides (E), rip
four ⅜"-thick strips from the
edge of ¾" stock 19⅞" long.
Using the Groove Detail in **4–12**
for reference, cut a ⅜" groove
³⁄₁₆" deep, centered along one
edge of each strip.

6 Dry-clamp two of the strips
together groove-to-groove.

*With the tables and spacers upside down, slide the all-thread rod through the
clamp blocks and guides to align the pieces. Nail the guides in place.*

4–14.

Slide a piece of ⅜" all-thread rod into the opening created by the mating grooves. The rod should fit snugly, yet slide back and forth in the opening. Enlarge the groove if necessary. Then, glue and clamp the two strips together to make each guide (E). Run the all-thread rod through the square opening in each laminated guide to remove any glue squeeze-out. Wait ten minutes and repeat the reaming process.

7 Sand each guide (E) smooth. Then, drill ¹⁄₁₆" pilot holes through each guide ³⁄₃₂" from the edges, where shown on **4–12**. Later, you'll drive nails through these holes to secure the guides to the bottom of the rear table (A).

8 Position the tables (A) and spacers (B, C) upside down on your workbench, with the spacers between the tables, as shown in **4–14**. Slide the all-thread rod through the clamp blocks (D) and guides (E) to align the pieces. Leave a ¼" gap at the front end of each guide (E) and next to one clamp block (D). Glue and nail the guides to just the rear table (A).

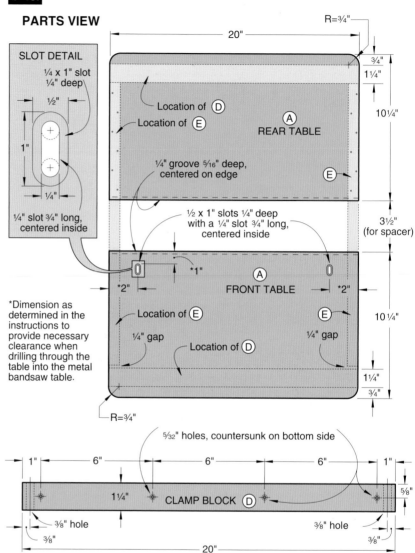

4–15.

PARTS VIEW

SLOT DETAIL
¼ x 1" slot
¼" deep
½"
1"
¼"
¼" slot ¾" long, centered inside

R=¾"
20"
¾"
1¼"
10¼"
Location of Ⓓ
Location of Ⓔ
Ⓐ
REAR TABLE
¼" groove ⁵⁄₁₆" deep, centered on edge
Ⓔ
3½" (for spacer)

½ x 1" slots ¼" deep with a ¼" slot ¾" long, centered inside

*Dimension as determined in the instructions to provide necessary clearance when drilling through the table into the metal bandsaw table.

*1"
*2"
Ⓐ
FRONT TABLE
*2"
Location of Ⓔ
¼" gap
Location of Ⓓ
Ⓔ
¼" gap
10¼"
1¼"
¾"
R=¾"

⁵⁄₃₂" holes, countersunk on bottom side
1"
6"
6"
6"
1"
1¼"
CLAMP BLOCK Ⓓ
⅝"
⅜" hole
⅜"
⅜" hole
⅜"
20"

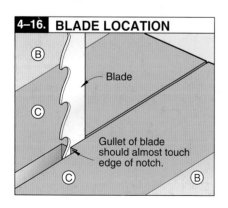

4–16. **BLADE LOCATION**
Ⓑ
Blade
Ⓒ
Gullet of blade should almost touch edge of notch.
Ⓒ
Ⓑ

Set the Blade Guides

1 Using a pair of panhead machine screws, attach the jig to your bandsaw. Adjust the wood tables and guides so the blade is positioned between the blade guides, as shown on **4–16**. The guides should be just next to the blade without touching it.

2 Change blades and reconnect the rear table. For changing the blade, simply remove the nuts and washers from the *front* end of the front table, and slide the *rear* table (A) off the metal bandsaw table.

MULTI-JIG CIRCLE-CUTTING GUIDE

This handy attachment (**4–17**) offers a breakthrough in circle cutting. Using the guide prevents start and stop marks that need cleaning up later. And it's as simple as a twist of the wrist with our sliding trammel and taped-in-place trammel disc. To make it, do the following:

1 Crosscut two pieces of ⅛ × ¾" flat steel to 18" long. Cut two pieces to ⅛ × 1 × 1⁷⁄₁₆" for the stop. See the Circle-Cutting Guide drawing in **4–18** for reference. Drill and tap the holes in the stop pieces, where shown on the drawing.

2 Cut the arms (F) to size from ½" solid stock. (We used birch stock.)

4–17.

3 Cut the arm spacers (G) and sliding trammel (H) to size plus 12" in length from ½" stock.

4 Cut a ¼" rabbet ¼" deep along the bottom outside edge of each arm (F), where shown in **4–18**. Test-fit the newly created tenon on each arm into the mating slots in the table (A).

5 Cut a ⅛" groove ⅜" deep, centered along the inside edge of each arm (F) and along the extra-long blanks for the arm spacers (G) and sliding trammel (H). Check that the ⅛ × ¾ × 18" flat steel stock slides smoothly without slop in the grooves in parts F, G, and H. Then, crosscut parts G and H to length from the 12"-long blanks.

6 Drill a ⅜" hole, centered, in the sliding trammel (H). Drive a ¼" threaded insert square into the sliding trammel.

7 Hacksaw the heads off two ¼" hexhead bolts with smooth upper shanks. Cut to ¾" long to form the two ¾"-long trammel points like those shown on the Sliding-Trammel Detail in **4–18**.

8 Assemble (dry-fit) the flat steel into the groove in the arms (F), and position parts G and H between the steel stock.

4–18.

CIRCLE-CUTTING GUIDE

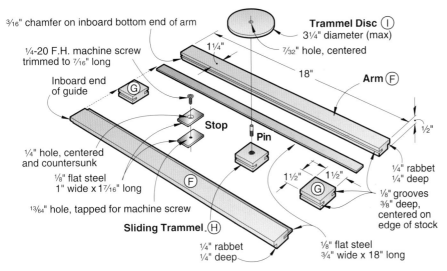

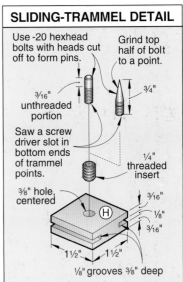

SLIDING-TRAMMEL DETAIL

Materials List for Multi-Jig Circle-Cutting Guide

PART	FINISHED SIZE			MTL.	QTY.
	T	W	L		
F arms	½"	1¼"	18"	B	2
G* arm spacers	½"	1½"	1½"	B	2
H* sliding trammel	½"	1½"	1½"	B	1
I trammel disc	¼"	3¼" dia.		H	1

*Cut parts oversize in length. Trim to finished size according to the instructions.

Materials Key: B = birch; H = hardboard.
Supplies: ⅛ x ¾" flat steel, 36" long; ⅛ x 1" flat steel, 3½" long; ¼" threaded insert; two ¼–20 x 2" hexhead bolts (for trammel points); ¼–20 x ½" flathead machine screw (trimmed to ⁷⁄₁₆" long).

The trammel (H) must slide back and forth on the flat steel. Sand the grooves in the trammel if necessary until it slides freely.

9 Epoxy the steel stock into the grooves in the arms (F). Immediately remove any excess epoxy. Later, position the spacers (G) and sliding trammel (H) between the two assemblies, keeping the ends of the spacers flush with the end of the arms. Epoxy the spacers in place; the trammel must be left free to slide on the steel stock. Rub a bit of paraffin on the bars if necessary so the trammel slides smoothly.

10 Using the ¼–20 machine screw, secure the stop to the steel stock so the stop will slide on the flat steel stock.

11 Cut the trammel disc (I) to shape, and drill a ⁷⁄₃₂" hole in its center.

Using the Circle-Cutting Guide

STEP 1

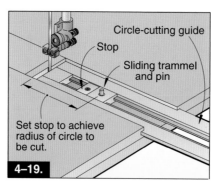

Set stop to achieve radius of circle to be cut.

4–19.

Remove the right-hand spacer guide, and replace it with the circle-cutting guide (**4–19**). Position and secure the sliding trammel so the distance from the center of the pin to the blade is equal to the radius of the circle you want to cut.

STEP 2

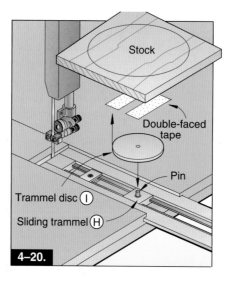

Stock

Double-faced tape

Pin

Trammel disc (I)

Sliding trammel (H)

4–20.

Using double-faced tape, adhere the trammel disc to the bottom center of the stock. Position it onto the pin (**4–20**).

STEP 3

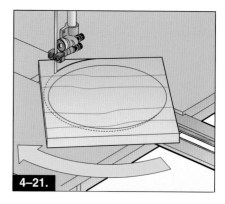

4–21.

Start with the edge of the stock against the blade (**4–21**). Turn the saw on, and slowly rotate the stock into the blade. At the same time, slowly push the stock (mounted to the disc and sliding trammel) toward the blade until the sliding trammel comes in contact with the stop. The blade will begin to cut a perfect circle at this point.

ARCH-CUTTING MADE EASY

Here is another way to cut arches with a bandsaw (**4–22** and **4–23**). An auxiliary table fastens to the bandsaw table by way of metal channel. (Clamps would work, too.) The left edge of the jig must be 5⁷⁄₁₆" from the left edge of the bandsaw table to cut correctly. A paddle outfitted with a toggle clamp holds the stock to be cut and rotates it through the blade. Here a sharp ¼" blade is the key to cutting success. See **4–23** for how the jig goes together.

4-22.

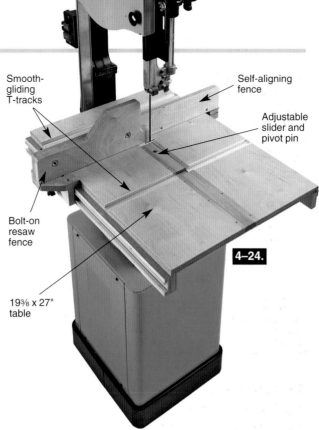

Smooth-gliding T-tracks

Self-aligning fence

Adjustable slider and pivot pin

Bolt-on resaw fence

19⅜ x 27" table

4-24.

4-23. ARCH-CUTTING JIG

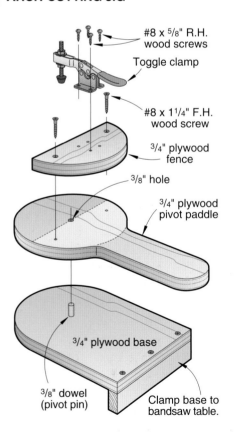

#8 x ⅝" R.H. wood screws

Toggle clamp

#8 x 1¼" F.H. wood screw

¾" plywood fence

⅜" hole

¾" plywood pivot paddle

¾" plywood base

⅜" dowel (pivot pin)

Clamp base to bandsaw table.

ULTIMATE SHOP-MADE TABLE SYSTEM

This fixture (**4-24**) does it all! Rip stock to width with a solid, self-aligning fence. Cut circles up to 32" in diameter. Crosscut with the miter gauge in one of two T-tracks. Resaw veneers using the bolt-on resaw fence. A 19⅜ x 27" table safely supports large workpieces. Plus, you change blades by removing only the slider strip, not the table. And beginning on *page 117* you'll find plans for three accessories that add maximum versatility: a tapering jig (**4-25**), a duplicating jig (**4-26**), and a feather board/single-point fence (**4-27**).

Want to take your bandsaw to a whole new level of performance and versatility? Here's your chance. After you build the jigs, see *page 120* for tips on putting the system to work in your shop.

Note: *You won't find the hardware listed in the supplies section due to the number of pieces and sources. See the Buying Guide on* page 115 *for all the hardware you need.*

4–25.

TAPERING JIG

Cut consistent tapers on parts such as table legs with ease.

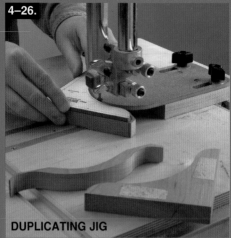

4–26.

DUPLICATING JIG

Produce identical curved parts quickly and accurately.

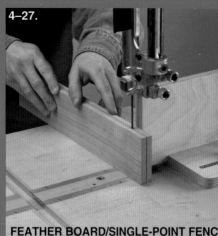

4–27.

FEATHER BOARD/SINGLE-POINT FENC

Resaw thick and narrow stock with this dual-function jig.

Start with the Table

1 Cut the table (A) to the size listed in the Materials List on *page 115*. On the table's top, lay out the 1⅜ × 18⁵⁄₁₆" slot, where dimensioned on **4–29**. Jigsaw the slot to within ¹⁄₁₆" of the lines.

2 Chuck a top-bearing pattern bit with a 1" cutter length in your router. Align a straightedge with one of the slot's long layout lines, and clamp it to the table (A). Next, rout along the edge of the slot. Repeat the process to trim the slot's end and other long edge.

SHOP TIP

How to Determine If the Jig will Fit Your Bandsaw

Take four measurements on your bandsaw, where shown on 4–28. The table will fit your saw provided your measurements stay within the limits shown.

BANDSAW TABLE
(Top view)

3 9¼" maximum — Blade

1 8¾" maximum

4 10½" minimum — Table slot

2 8¾" maximum

Bandsaw frame

4–28.

4–29. **TABLE**

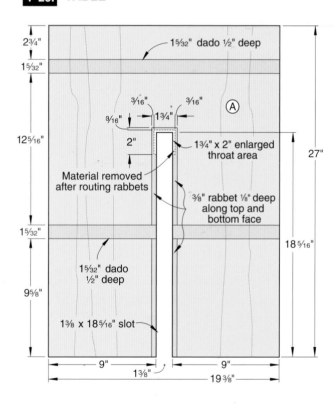

2¾"

1⁵⁄₃₂"

1⁵⁄₃₂" dado ½" deep

³⁄₁₆"   ³⁄₁₆"

³⁄₁₆"   1¾"

(A)

12⁵⁄₁₆"

2"

1¾" x 2" enlarged throat area

27"

Material removed after routing rabbets

⅜" rabbet ⅛" deep along top and bottom face

1⁵⁄₃₂"

18⁵⁄₁₆"

1⁵⁄₃₂" dado ½" deep

9⅝"

1⅜ x 18⁵⁄₁₆" slot

9"   1⅜"   9"

19⅜"

3 Refit your router with a ⅜" rabbeting bit. Rout a ⅜" rabbet ⅛" deep along the slot's edges on both faces of the table, where shown on **4–29**, to receive the ⅛"-thick aluminum bars, where shown on **4–30**. Square the rabbet corners with a chisel.

4–30. EXPLODED VIEW

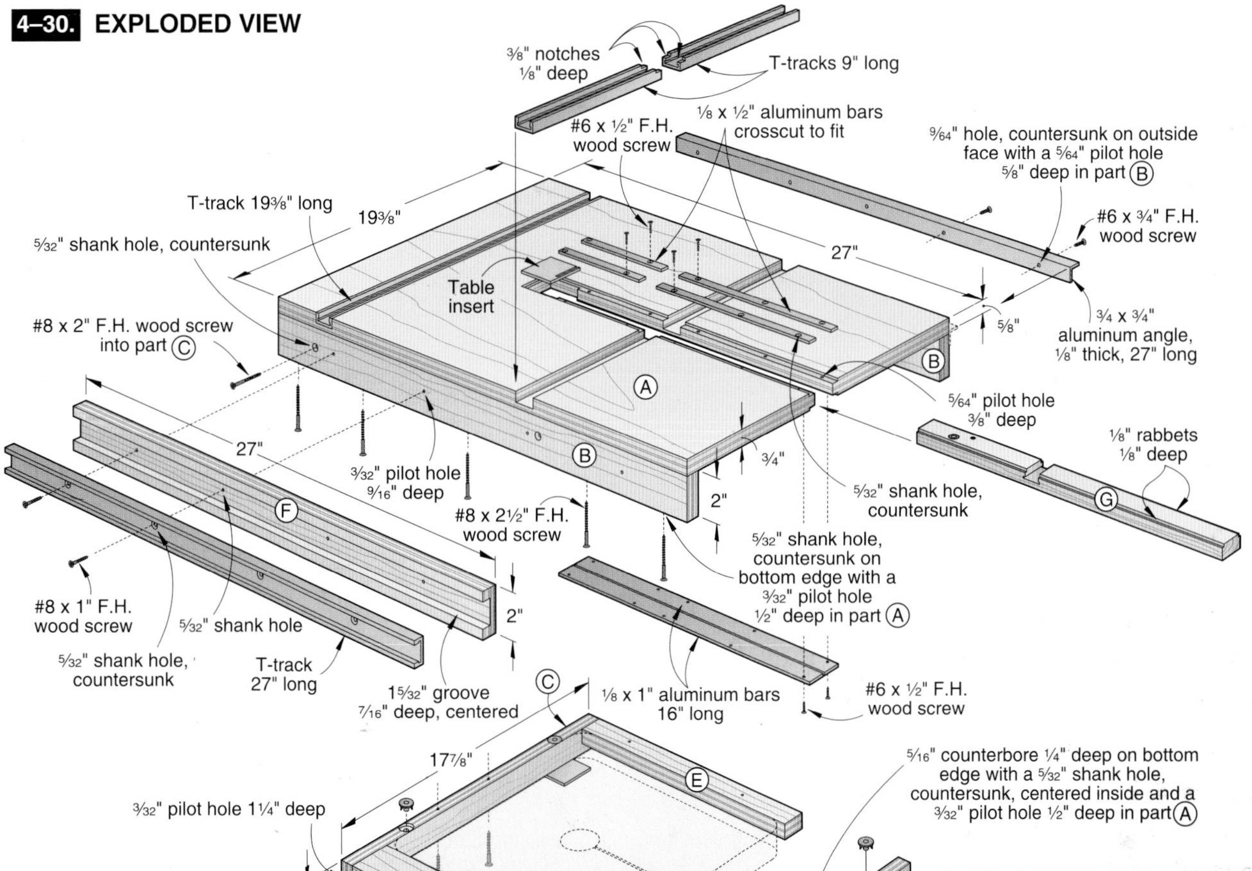
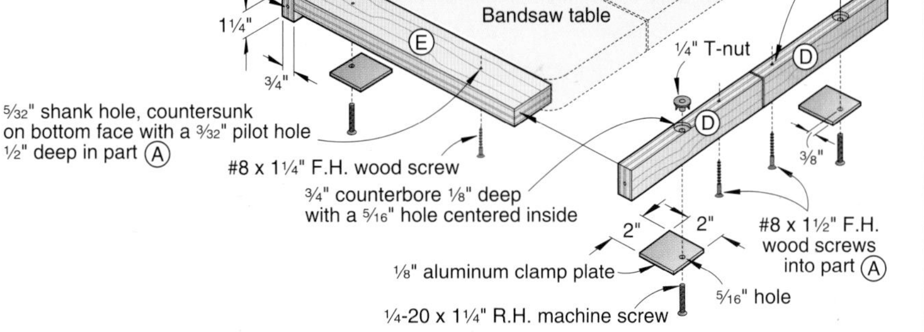
3/8" notches
1/8" deep
T-tracks 9" long
#6 x 1/2" F.H. wood screw
1/8 x 1/2" aluminum bars crosscut to fit
9/64" hole, countersunk on outside face with a 5/64" pilot hole 5/8" deep in part B
T-track 19 3/8" long
19 3/8"
5/32" shank hole, countersunk
#6 x 3/4" F.H. wood screw
27"
Table insert
#8 x 2" F.H. wood screw into part C
5/8"
3/4 x 3/4" aluminum angle, 1/8" thick, 27" long
B
A
5/64" pilot hole 3/8" deep
1/8" rabbets 1/8" deep
27"
B
3/4"
3/32" pilot hole 9/16" deep
G
5/32" shank hole, countersunk
F
2"
#8 x 2 1/2" F.H. wood screw
5/32" shank hole, countersunk on bottom edge with a 3/32" pilot hole 1/2" deep in part A
2"
#8 x 1" F.H. wood screw
5/32" shank hole
5/32" shank hole, countersunk
T-track 27" long
1 5/32" groove 7/16" deep, centered
C
1/8 x 1" aluminum bars 16" long
#6 x 1/2" F.H. wood screw
17 7/8"
E
5/16" counterbore 1/4" deep on bottom edge with a 5/32" shank hole, countersunk, centered inside and a 3/32" pilot hole 1/2" deep in part A
3/32" pilot hole 1 1/4" deep
1 1/4"
Bandsaw table
E
1/4" T-nut
D
3/4"
5/32" shank hole, countersunk on bottom face with a 3/32" pilot hole 1/2" deep in part A
D
3/8"
#8 x 1 1/4" F.H. wood screw
3/4" counterbore 1/8" deep with a 5/16" hole centered inside
2"
2"
#8 x 1 1/2" F.H. wood screws into part A
1/8" aluminum clamp plate
5/16" hole
1/4-20 x 1 1/4" R.H. machine screw

4–31. TABLE INSERT

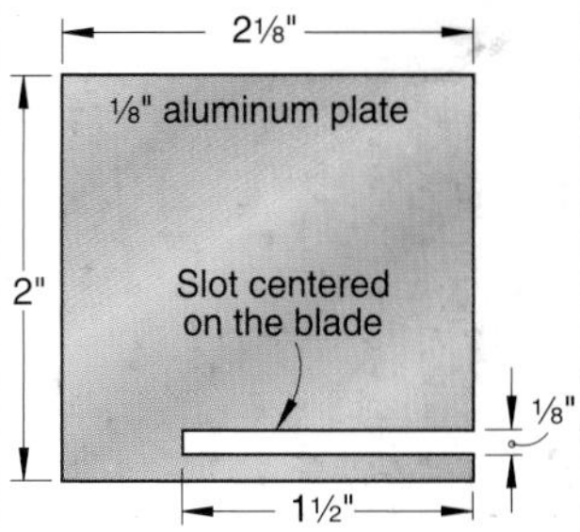
2 1/8"
1/8" aluminum plate
2"
Slot centered on the blade
1/8"
1 1/2"

Note: *For safe operation when using the system's jigs, ensure that all of the aluminum parts fit flush with the table's top surface.*

4 Lay out the 1 3/4 x 2" enlarged throat area at the end of the slot, where dimensioned on **4–29**. (This provides clearance to facilitate blade changes.) Jigsaw the opening to shape.

5 Using a dado blade in your tablesaw, cut the 1 5/32" dadoes 1/2" deep in the table's top, where dimensioned, to receive the T-tracks shown on **4–30**.

Cut the Aluminum Parts

1 From a 48"-long piece of 1 5/32"-wide aluminum T-track, hacksaw a 19 3/8"-long piece and two 9"-long pieces to fit the exact length of the table's dadoes, where shown on **4–30**. Cut the 3/8" notches 1/8" deep in the inboard ends of the 9"-long pieces, where shown. Position the notched pieces in the dadoes.

2 From a 1/8 x 2 x 12" aluminum bar, cut four 2"-long pieces

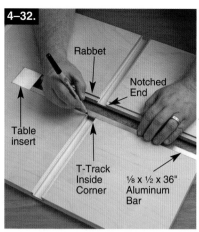

4–32.

Using a fine-tip permanent marker, mark the bar exactly at the inside corners of the T-track and at the table's end, and crosscut to length.

for the table clamp plates and a 2⅛"-long piece for the table insert, where shown on **4–30** and **4–31**. Drill a ⁵⁄₁₆" hole through each clamp plate, where dimensioned. Set the plates aside. Place the insert in the table. (You'll cut the slot in the insert later.)

3 Position a ⅛ x ½ x 36" aluminum bar in the rabbet along one edge of the slot on the top of the table (A), flush against the table insert, as shown in **4–32**. Mark the bar for the exact lengths of the pieces, where shown on **4–30**, so they'll fit flush with the inside corners of the T-track. Crosscut the pieces to length. Repeat this process on the opposite side of the slot.

4 Drill countersunk shank holes through the four bars, where shown. Position the bars in the rabbets. Using the holes as guides, drill pilot holes in the table. Remove the bars and table insert.

5 From a ⅛ x 1 x 36" aluminum bar, hacksaw two 16"-long pieces for the rabbets on the bottom of the table. Position the pieces in the rabbets, flush with the end of the table. Mark the locations for the screw holes, making sure they do not line up with the screw holes for the top bars. Drill countersunk shank holes through the bars. Place the bars in position, and drill the pilot holes. Remove the bars.

6 From a 32"-long piece of T-track, cut a 27"-long piece for the fence rail (F), shown on **4–30**. From a 36"-long piece of ¾ x ¾" aluminum angle ⅛" thick, cut a 27"-long piece for the back fence rail. Set these pieces aside.

7 Sand the table (A) to 180 grit. Abrade the bottom and sides of the three T-track pieces for the table's top with 40-grit sandpaper. Remove the dust. Apply five-minute epoxy along the bottom and sides of the table's dadoes. Install and clamp the T-tracks in the dadoes. Screw the aluminum bars in their rabbets. Set the table aside while the epoxy cures.

Complete the Table

1 Cut the front and back rails (B) to the sizes listed. Position the rails on the bottom of the table (A), where shown on **4–30**. Drill mounting holes through the bottom edge of each rail. Glue and screw the rails to the table.

2 Remove the round insert from your bandsaw table. Place the plywood bandsaw table assembly (A/B) on your bandsaw table. Align the plywood table's ⅛" slot between its bottom aluminum bars with the bandsaw table's slot. Locate the end of the aluminum bars in the throat area ³⁄₁₆" from the blade, as shown in **4–33**. Insert a piece of ⅛" hardboard between the tables' slots, as shown, to keep them aligned. On the plywood table's bottom, scribe along the band saw table's sides.

3 Measure the thickness of your bandsaw table. (Ours measured 1¼".) Cut the side rail (C) and split side rails (D) to the lengths listed and width equal to

Below left: With the plywood table correctly positioned on the bandsaw table, insert a ⅛" hardboard scrap between the tables' aligned slots.
Below right: Find the width for the fillers (E) by measuring the space between the front and back rails (B) and the bandsaw table.

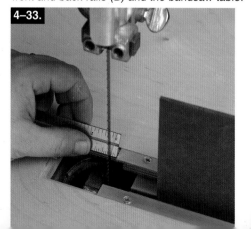

4–33.

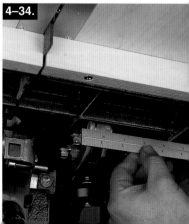

4–34.

your measured table thickness. Position the side rail (C) under the plywood table, tight against the side of the bandsaw table, where shown on **4–30**. Make a mark across the bottom edge of the rail 1" in from the front and back edges of the bandsaw table. These marks locate the centerlines for the clamp-plate screw holes. Repeat the process to mark the split side rails (D), holding them tight against the front and back rails (B). Remove the plywood table.

4 At the marked centerlines on the rails (C, D), drill a ¾" counterbore ⅛" deep with a ⁵⁄₁₆" hole centered inside. Install a ¼" T-nut in each counterbore.

5 With the plywood table bottom side up, place the side rail (C) and split side rails (D) in position, aligning their inside edges with the scribe marks. Drill counterbored mounting holes through the rails, where shown. Drive the screws. Use a screw length appropriate for the width of your rails. Note that if the rails directly align with the T-tracks in the table's top, you'll need to attach them from the top by drilling countersunk shank holes through the T-tracks.

6 Drill pilot and countersunk shank holes through the front and back rails (B) into the side rails (C, D), where shown on **4–30**. Drive the screws.

7 Reinstall and align the ply-wood table on the bandsaw table. Measure between the side rail (C) and split side rails (D) for the length of the fillers (E). Measure for their widths, as shown in **4–34**. (The widths may be different.) Cut the pieces to size. Place each filler in position. Drill mounting holes through the fillers, where shown. Drive the screws.

8 Cut the fence rail (F) to size. Cut a 1⁵⁄₃₂" groove ⁷⁄₁₆" deep centered along the front face of the rail to receive the 27"-long piece of T-track. Drill counter-sunk shank holes through the T-track. Place the T-track in the rail's groove. (The T-track sits ¹⁄₁₆" proud of the rail's face.) Position the rail/track assembly against the front rail (B) with the bottom edges flush. Using the holes in the T-track as guides, drill mounting holes through the fence rail and into the front rail. Glue and screw the assembly to the front rail.

9 Retrieve the aluminum-angle back fence rail. Drill counter-sunk shank holes, where shown. Position it on the back rail (B) ¾" below the top of the plywood table, where dimensioned on **4–30**. Drill pilot holes in the back rail. Screw it in place.

10 Apply two coats of satin polyurethane to the completed plywood table. When dry, fit the table on the bandsaw. Screw the aluminum clamp plates to the side rails (C, D) with ¼–20 roundhead machine screws. You may need to use a different length screw than shown to suit your rails' width.

Add the Slider

1 To cut circles with your system, you need to make the slider (G). Cut the part to size. Chuck a rabbeting bit in your table-mounted router. Cut a ⅛" rabbet ⅛" deep along the top edges of the slider, where shown on **4–35**.

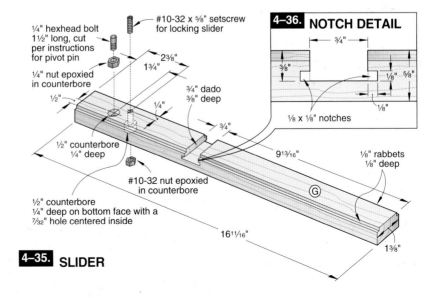

4–36. NOTCH DETAIL

¾"

⅜"

⅛" · ⅝"

⅛"

⅛ x ⅛" notches

¼" hexhead bolt 1½" long, cut per instructions for pivot pin

#10-32 x ⅝" setscrew for locking slider

¼" nut epoxied in counterbore

2⅜"

1¾"

½"

¼"

¾" dado ⅜" deep

¾"

½" counterbore ¼" deep

#10-32 nut epoxied in counterbore

9¹³⁄₁₆"

⅛" rabbets ⅛" deep

½" counterbore ¼" deep on bottom face with a ⁷⁄₃₂" hole centered inside

G

16¹¹⁄₁₆"

1⅜"

4–35. SLIDER

2 Insert the slider in the plywood table's slot, flush with its right end. Scribe along the inboard ends of the ⅛ x ½" aluminum bars to mark the location for the slider's dado. Remove the slider. Cut the ¾" dado ⅜" deep. Bandsaw, with the slider resting on its edge, the notches in the sides of the dado at the bottom, where dimensioned on **4–36**.

3 Mark the centerpoints for the ½" counterbores on the top and bottom of the slider, where dimensioned on **4–35**. Note that the center of the top counterbore and the front of the bandsaw blade must be the same distance

from the plywood table's front edge for proper circle-cutting operation. (You may need to adjust the counterbore's location from the dimension shown to suit your saw.) Using a ½" Forstner bit, drill the counterbores ¼" deep at the centerpoints. Then, drill a 7/32" hole centered inside the bottom counterbore.

4 Epoxy a ¼" nut in the top counterbore and a #10-32 nut in the bottom counterbore. When the epoxy cures, sand the slider. Apply two coats of finish.

5 Thread a #10-32 x⅝" setscrew (for locking the slider) into the #10-32 nut. To

make a pivot pin, mark a ½" length on a ¼" hexhead bolt 1½" long that includes ¼" of thread and ¼" of smooth shank. Cut the length from the bolt, and file its ends and edges smooth. Thread it into the ¼" nut. Now, insert the slider in the table.

Time for the Fences

1 Cut the fence (H) and stiffener (I) to size. Bandsaw and sand the ½" radii on the fence, where shown on **4–37**, and the stiffener, where shown on **4–40**. Mark the angled end on the stiffener, where dimensioned, and cut it to shape.

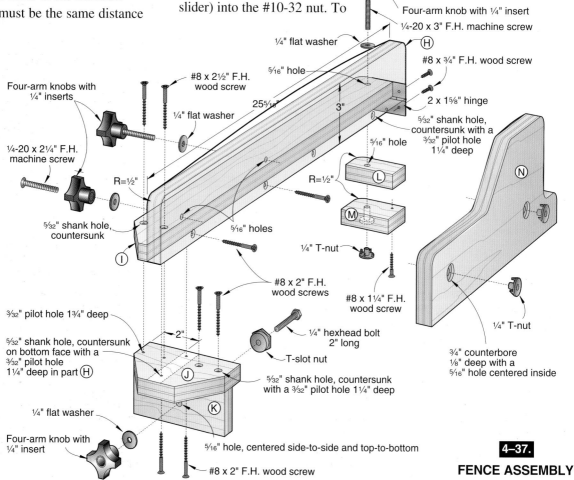

Four-arm knob with ¼" insert
¼-20 x 3" F.H. machine screw
¼" flat washer
5/16" hole
#8 x 2½" F.H. wood screw
25 5/16"
3"
(H)
#8 x ¾" F.H. wood screw
2 x 1⅝" hinge
5/32" shank hole, countersunk with a 3/32" pilot hole 1¼" deep
5/16" hole
(L)
(N)
Four-arm knobs with ¼" inserts
¼" flat washer
¼-20 x 2¼" F.H. machine screw
R=½"
R=½"
(M)
¼" T-nut
5/32" shank hole, countersunk
(I)
5/16" holes
#8 x 2" F.H. wood screws
#8 x 1¼" F.H. wood screw
¼" T-nut
¾" counterbore ⅛" deep with a 5/16" hole centered inside
3/32" pilot hole 1¾" deep
5/32" shank hole, countersunk on bottom face with a 3/32" pilot hole 1¼" deep in part (H)
2"
(J)
¼" hexhead bolt 2" long
T-slot nut
5/32" shank hole, countersunk with a 3/32" pilot hole 1¼" deep
¼" flat washer
(K)
Four-arm knob with ¼" insert
5/16" hole, centered side-to-side and top-to-bottom
#8 x 2" F.H. wood screw

4–37.

FENCE ASSEMBLY

4–38. HINGE DETAIL
(Viewed from back of fence)

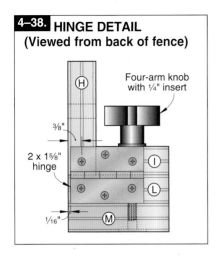

Sand the parts smooth. Do not drill the holes in them yet.

2 Position the stiffener against the fence, where shown on **4–37**. Drill mounting holes, where, shown. Glue and screw parts H and I together.

3 Cut the fence plate (J) to size. Lay out and cut the plate's angled sides, where dimensioned on **4–40**.

4 Position the fence/stiffener assembly (H/I) on the fence plate (J), where shown on **4–37**. Square the fence to the plate's back edge. Clamp the assembly together. Drill two mounting holes through the bottom of the plate into the fence, where shown. Drive the screws.

5 Cut the clamp plate (K) to size. Drill a 5/32" hole, centered side-to-side and top-to-bottom, through the part. Set the fence assembly (H/I/J) on the plywood table with the back edge of the fence plate (J) flush against the table's front edge. Position the

clamp plate under the fence plate, flush against the T-track in the fence rail (F). Clamp the plates together. Drill mounting holes through the top of the stiffener (I) and fence plate, where shown. Glue and screw the plates together. Insert a 1/4" hexhead bolt 2" long with a T-slot nut through the hole in the clamp plate from the rear. Secure the plates with a 1/4" flat washer and four-arm knob having a 1/4" insert, as shown in **4–37**.

6 Cut the rear top and bottom clamps (L, M) to size. Place clamp L on top of clamp M, aligning their back edges. Drill a

Cutting Diagram

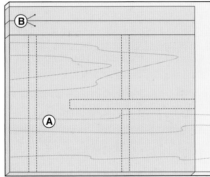

3/4 x 24 x 30" Baltic birch plywood

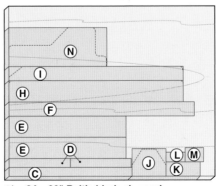

3/4 x 24 x 30" Baltic birch plywood

Ⓖ

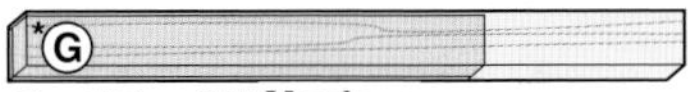

3/4 x 1½ x 24" Maple
*Plane or resaw to the thickness listed in the Materials List.

pilot and countersunk shank hole through the bottom clamp, where shown on **4–40**. Glue and screw the parts together. When the glue dries, mark the 1/2" radius on the assembly. Bandsaw and sand the radius to shape.

Materials List for Ultimate Table System

		FINISHED SIZE				
		T	W	L	MTL..	QTY.
BANDSAW TABLE						
A	table	3/4"	19⅜"	27"	BB	1
B	front and back rails	3/4"	2"	27"	BB	2
C	side rail	3/4"	+	17⅞"	BB	1
D	split side rails	3/4"	+	8⅞"	BB	2
E	fillers	3/4"	+	+	BB	2
F	fence rail	3/4"	2"	27"	BB	1
G	slider	5/8"	1⅜"	16¹¹⁄₁₆"	M	1
FENCE ASSEMBLY						
H	fence	3/4"	3"	25⁵⁄₁₆"	BB	1
I	stiffener	3/4"	2"	25⁵⁄₁₆"	BB	1
J	fence plate	3/4"	3¹³⁄₁₆"	5"	BB	1
K	clamp plate	3/4"	2"	5"	BB	1
L	rear top clamp	3/4"	1⅜"	2¾"	BB	1
M	rear bottom clamp	3/4"	2"	2¾"	BB	1
N	resaw fence	3/4"	5½"	14¼"	BB	1

+Dimensions determined by your bandsaw measurements. See the instructions.

Materials Key: BB = Baltic birch plywood; M = maple.
Supplies: Five-minute epoxy;
Bits and Blades: Dado-blade set; top bearing pattern bit with 1" long cutter; 1/2" Forstner bit; 1/8" and 3/8" rabbeting bits.
Buying Guide: Hardware kit for ultimate bandsaw table. Contains all hardware (screws included) required for one table. Order kit no. UBT, about $40 ppd., from Schlabaugh and Sons Woodworking. Call 800/346-9663, or go to www.schsons.com to order. Master hardware kit for ultimate bandsaw table and accessories. Contains all hardware (screws included) required for one table and all accessories. Order kit no. MAS-BAN, about $75 ppd. Address and telephone above.

This Stud's for You

Finding knobs with custom-length threaded studs can be a challenge. Here's an easy way to make your own. Thread a ¼–20 flathead machine screw of sufficient length (cut if needed) completely into the knob. Mark the threads immediately below the knob with a permanent marker. Back out the screw until you see the mark. Apply five-minute epoxy to the threads above the mark. Then the epoxied screw in the knob, and wipe off any squeeze-out.

4–39. Apply epoxy to threads above mark.

Epoxy

7 Position the clamp assembly (L/M) at the rear of the fence assembly, where shown on **4–37**, aligning their back edges and ½" radii. Clamp the parts together. Screw a 2 × 1⅝" hinge to the back of the fence and clamp assembly, where shown on **4–38**. Drill a ¾" counterbore ⅛" deep in the bottom clamp (M), where dimensioned on **4–40**. Now, drill a ⁵⁄₁₆" hole centered inside through the clamp assembly and stiffener (I), where shown on **4–37** and **4–40**. Install a ¼" T-nut in the counterbore.

8 Make a four-arm knob with a ¼–20 × 3" threaded stud for the rear clamp assembly. (See the Shop Tip "This Stud's For You," *above*.) Install the knob with a ¼" flat washer in the ⁵⁄₁₆" hole in the stiffener (I), where shown.

9 Cut the resaw fence (N) to size. Lay out its contour, where dimensioned on **4–40**. Bandsaw the fence to shape, and sand it smooth. Drill the needed holes, where dimensioned. Position the resaw fence against the right side of fence H, with their front ends and bottom edges aligned. With a backer board placed against the left side of the fence to prevent tear-out and, using the centered holes in the counterbores as guides, drill ⁵⁄₁₆" holes through the fence. Apply two coats of finish to the fences.

10 Install ¼" T-nuts in the resaw fence's counterbores. Epoxy two ¼–20 × 2¼" flathead machine screws in two four-arm knobs. Secure the resaw fence to the fence with the knobs and ¼" flat washers.

11 Finally, place the aluminum table insert in the table opening against the blade's cutting edge. Referring to **4–31**, on *page 111*, mark a ⅛" slot 1 ½" long, centered on the blade, on the plate. Hacksaw or scroll-saw the slot, and install the plate. Now, it's time to put this awesome system to work.

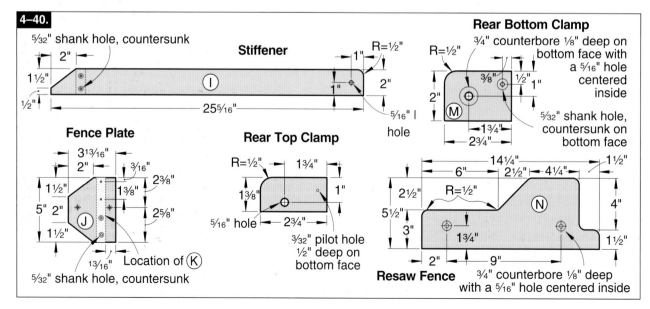

4–40.

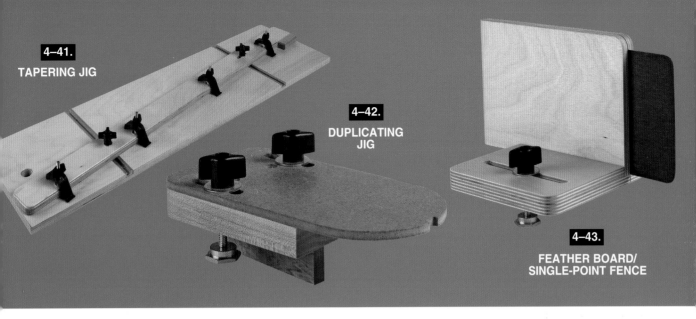

4–41.
TAPERING JIG

4–42.
DUPLICATING JIG

4–43.
**FEATHER BOARD/
SINGLE-POINT FENCE**

Three Accessories for the Table System

Add even more operation options with this trio of hardworking jigs (**4–41** to **4–43**).

Accessory 1:
TAPERING JIG (4–44)

Start with the Sled

1 Cut the sled (A) to size from Baltic birch plywood. Drill a ¾" hanging hole, where shown on **4–46**. Bandsaw and sand ½" radii on two corners, and rout ⅛" chamfers along one side and both ends of both the sled's faces and around the hanging hole, where shown. Cut dadoes, where dimensioned, to accept the T-tracks.

2 Measure the distance from the side of your bandsaw's blade to the edge of the miter-gauge slot in which the jig's miter-guide bar will run. Cut a ¾" groove ¹⁄₁₆" deep at this location in the sled's bottom face. Resaw and plane the guide bar (B) to size, and shape its end, as shown on **4–47**. Clamp the guide bar in the groove. Drill

pilot and countersunk shank holes through the guide bar and into the sled. Drive the screws.

Note: Do not glue the guide bar in place. By removing the guide bar, you also can use the taper jig on your tablesaw, using the saw's fence to guide the jig.

3 Cut two pieces of aluminum T-track to the length shown. Mix quick-setting epoxy, and epoxy and clamp the tracks into the dadoes. To prevent the possibility of the blade coming in

contact with the tracks, position the tracks flush with the sled's chamfered edge.

Add the Fence

1 Cut the fence (C) to size. Mark the centers of the four counterbored holes for the hold-down clamps on the bottom of the fence. Drill the counterbores and holes. Mark and drill a ¼" hole for the first clamp knob. Drill two ⁵⁄₁₆" holes for the second clamp knob's slot. Form the slot, as shown in **4–45**. Cut the stop (D) to size, and glue and clamp it to

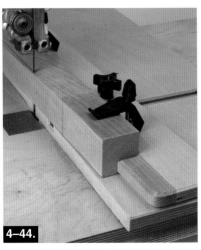

4–44.

The tapering jig in use.

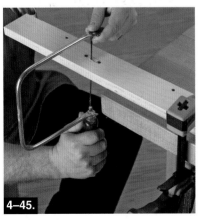

4–45.

Draw lines tangent to the two ⁵⁄₁₆" holes, and saw out the slot with a coping saw.

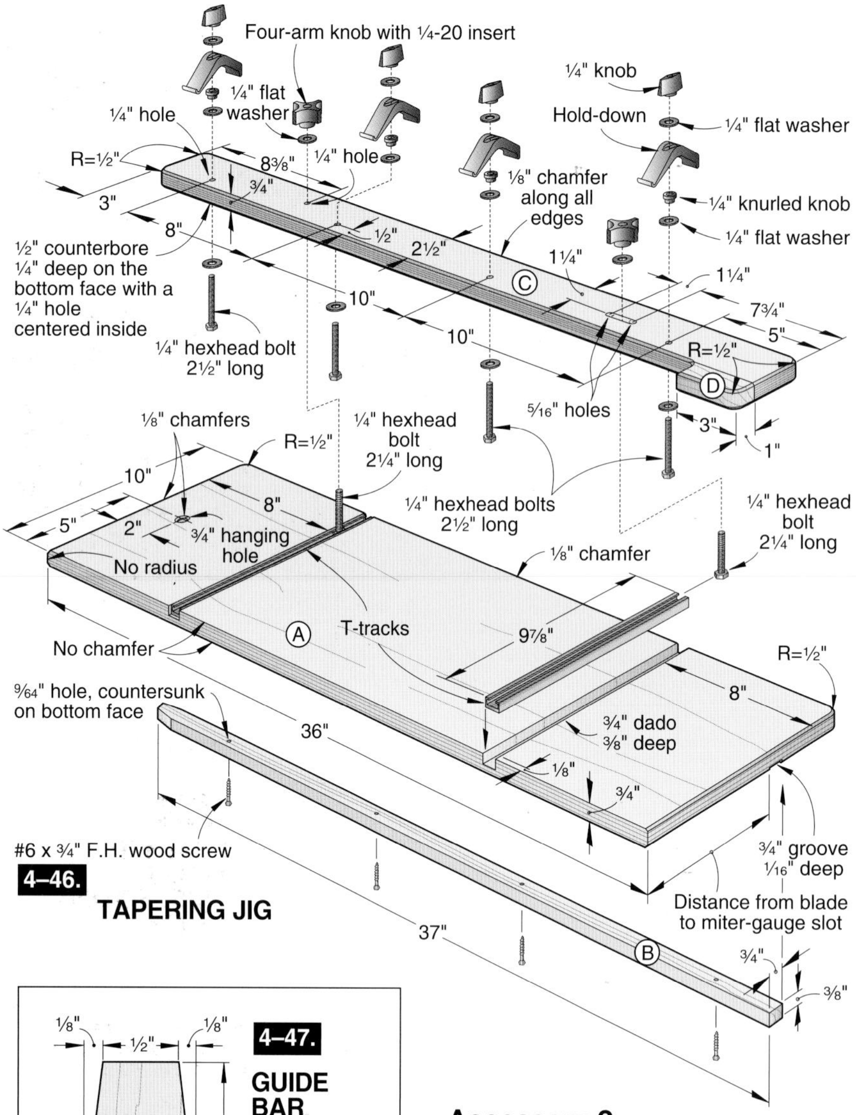

Four-arm knob with ¼-20 insert
¼" flat washer
¼" hole
¼" hole
R=½"
8⅜"
¾"
3"
8"
½"
2½"
½" counterbore ¼" deep on the bottom face with a ¼" hole centered inside
¼" hexhead bolt 2½" long
10"
10"
¼" knob
Hold-down
¼" flat washer
⅛" chamfer along all edges
¼" knurled knob
¼" flat washer
1¼"
1¼"
7¾"
5"
R=½"
5⁄16" holes
3"
1"
⅛" chamfers
R=½"
¼" hexhead bolt 2¼" long
10"
5"
2"
8"
¾" hanging hole
No radius
¼" hexhead bolts 2½" long
¼" hexhead bolt 2¼" long
⅛" chamfer
T-tracks
9⅞"
No chamfer
R=½"
8"
9⁄64" hole, countersunk on bottom face
36"
¾" dado ⅜" deep
⅛"
¾"
#6 x ¾" F.H. wood screw
¾" groove 1⁄16" deep
Distance from blade to miter-gauge slot
37"
¾"
⅜"

4–46. TAPERING JIG

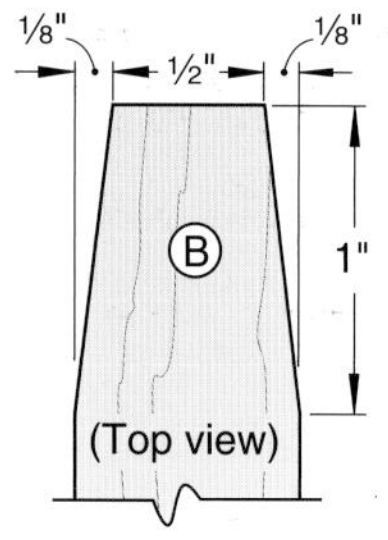

⅛"
½"
⅛"
1"
(Top view)

4–47. GUIDE BAR END

the fence. Bandsaw and sand ½" radii, where shown, and then rout ⅛" chamfers along all the top and bottom ends and edges.

2 Sand the sled and fence to 220 grit. Apply two coats of satin polyurethane to the parts. Install the hardware.

Accessory 2: DUPLICATING JIG (4–48 and 4–49)

1 Cut the guide (E) to size. Bandsaw and sand the 2" radius, where shown on **4–49**, and then file the blade notch. Cut the riser (F) from stock that is 1⁄16" thicker than the parts you will be sawing with the duplicating jig.

Note: You may want to make several duplicating jigs to accommodate common stock thicknesses.

4–48.

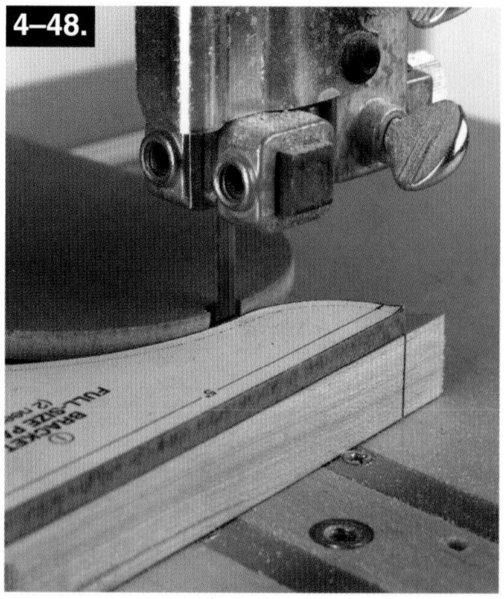

2 Glue and clamp the guide to the riser, keeping their ends and edges flush. Drill pilot and countersunk shank holes, and drive the screws. Drill ⅜" holes to form the ends of the slots. Saw them out in the same manner shown in **4–45**. Bandsaw and sand the ½" radii.

3 Apply polyurethane. Install the hardware.

4–49. DUPLICATING JIG

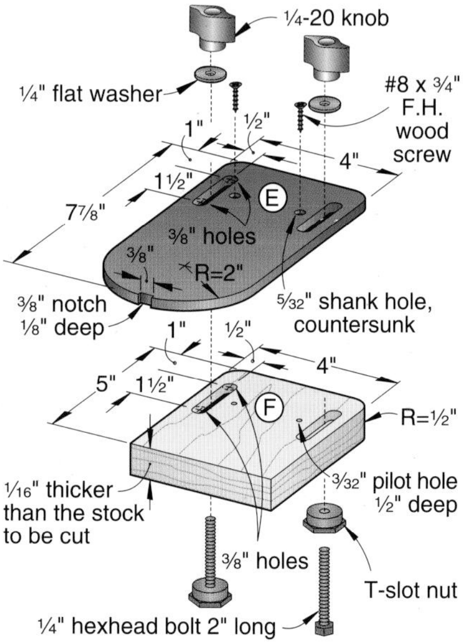

¼-20 knob
¼" flat washer
#8 x ¾" F.H. wood screw
1"
½"
4"
1½"
7⅞"
⅜" holes
⅜"
R=2"
⅜" notch ⅛" deep
5⁄32" shank hole, countersunk
1"
½"
4"
5"
1½"
R=½"
1⁄16" thicker than the stock to be cut
3⁄32" pilot hole ½" deep
⅜" holes
T-slot nut
¼" hexhead bolt 2" long

Accessory 3:
FEATHER BOARD/SINGLE-POINT FENCE
(4–50 and 4–51)

1 Cut the base (G) to size and plow the centered dado, where shown on **4–53**. Drill ⅜" holes to form the ends of the slots, and saw out the waste. Bandsaw and sand ½" radii on all four corners. Cut the upright (H) to size.

2 Rout a pair of ⅜" round-overs on one end, forming a full round. Bandsaw and sand the ½" radius, where shown, and then finish-sand the parts. Glue and clamp the upright into the base's dado with the feather board end flush with the base's edge. Drill pilot and countersunk shank holes through the base into the upright, and drive the screws.

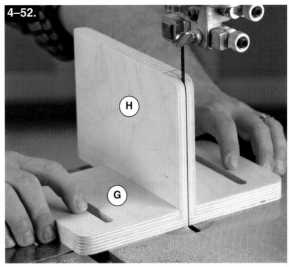

4–52.

With the base (G) and upright (H) assembled, bandsaw a centered kerf for the plastic-laminate feather.

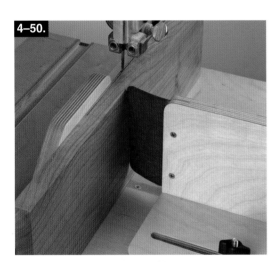

4–50.

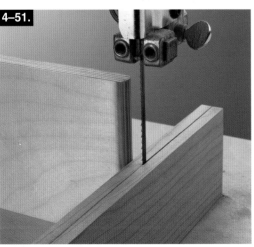

4–51.

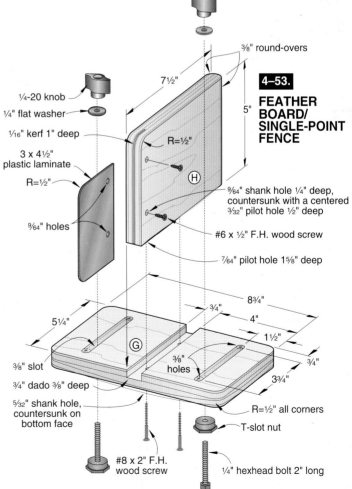

⅜" round-overs

4–53.
FEATHER BOARD/ SINGLE-POINT FENCE

¼-20 knob

¼" flat washer

1/16" kerf 1" deep

3 x 4½" plastic laminate

R=½"

9/64" holes

7½"

5"

R=½"

(H)

9/64" shank hole ¼" deep, countersunk with a centered 3/32" pilot hole ½" deep

#6 x ½" F.H. wood screw

7/64" pilot hole 1⅝" deep

8¾"

¾"

4"

1½"

¾"

3¾"

5¼"

(G)

⅜" holes

⅜" slot

¾" dado ⅜" deep

5/32" shank hole, countersunk on bottom face

R=½" all corners

T-slot nut

#8 x 2" F.H. wood screw

¼" hexhead bolt 2" long

3 Form the groove for the plastic-laminate feather, as shown in **4–52**. Cut a piece of laminate to the size shown, and sand the ½" radii. Insert the laminate into the kerf, aligning its bottom edge with the upright's bottom edge. Drill pilot and countersunk shank holes. Remove the laminate and apply two coats of satin polyurethane. With the finish dry, reinsert the laminate, drive the screws, and install the hardware.

Materials List for 3 Accessories

PART	FINISHED SIZE			MTL.	QTY.
	T	W	L		
TAPERING JIG					
A sled	¾"	10"	36"	BB	1
B guide bar	⅜"	¾"	37"	M	1
C fence	¾"	2½"	36"	BB	1
D stop	¾"	1"	3"	M	1
DUPLICATING JIG					
E guide	¼"	4"	7⅛"	H	1
F riser	†	4"	5"	M	1
FEATHER BOARD/SINGLE-POINT FENCE					
G base	¾"	5¼"	8¾"	BB	1
H upright	¾"	5"	7½"	BB	1

†Thickness varies. See the instructions
Materials Key: BB = Baltic birch plywood; M = maple; H = tempered hardboard.
Supplies: Quick-setting epoxy.
Blades and Bits: Stack dado set; chamfer and ⅜" round-over router bits.
Buying Guide
Hardware Kits: Kits contain all the hardware shown on the drawings, including the plastic laminate for the feather board. Tapering jig hardware kit no. BTJ, about $20 ppd.; duplicating jig hardware kit no. DUP, about $15 ppd.; feather board/single-point fence hardware kit no. FB-SPF, about $16 ppd. Schlabaugh & Sons Woodworking, 720 14th Street, Kalona, IA 52247. Call 800/346-9663. Or visit their Web site: www.schsons.com.

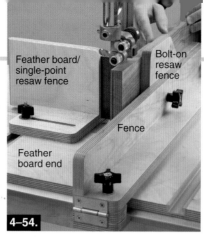

4–54.

You can slice veneers accurately and easily with this setup.

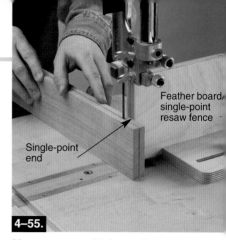

4–55.

You can resaw thicker and thinner stock easily with this setup.

Slice Veneers Accurately and Easily with this System

To resaw a thin slice (⅛" or less) from any board, as shown in **4–54**, you need rock-solid support on both sides of your stock to hold it in position vertically. This system's bolt-on resaw fence and feather board deliver, especially when you're working with stock more than about 5" wide, such as you'd use when creating your own veneer.

To set up the system for resawing, do the following:

1 Position the fence and lock it down with the front and rear knobs.

2 Slide the feather board end of the feather board/single-point resaw fence into the T-slot to support the outside face of the stock.

3 Lock in the feather board so the workpiece contacts the laminate just in front of the blade and flexes the laminate to provide support without binding. Make sure to crank up the blade tension to keep the blade taut, and push the workpiece slowly as you cut.

Keep the Workpiece Running True

When resawing thicker slabs (¼" or greater), or when working with narrow stock, vertical support isn't as important as keeping the workpiece running true. Bandsaw blades sometimes twist slightly as you push the workpiece, meaning you'll have to angle the workpiece slightly to compensate for this "drift." You can do this by guiding your stock against the bullnose end of the feather board/ single-point resaw fence, as shown in **4–55**.

To keep the workpiece running true, do the following:

1 Align the bullnose end with the blade's teeth, leaving a gap equal to the thickness of the piece you wish to cut.

2 Tighten the knobs to lock the fence in the T-slot.

3 Mark a layout line on the top edge of your workpiece, and push the piece through the cut while pressing one face against the bullnose. Note that the standard fence does not get used in this resawing technique.

Rip Stock Cleanly with this Standard Fence

The standard fence provides sturdy support for ripping boards to width, as shown in **4–56**, or other precise operations, such as cutting tenons. Just loosen the front and rear locking knobs, slide the fence into place, and then tighten the front knob. This locks the fence and automatically aligns it with the blade. Then lock the rear knob for surefooted cuts.

Duplicate Patterns Effortlessly

Need to cut multiple curved pieces to the exact same size and shape? This duplicating jig, shown in **4–57**, is just the ticket, with its hardboard guide that's notched to fit snugly around the blade. Attach a hardboard pattern, cut in the exact shape you require, to your workpiece using double-faced tape. Then just feed the workpiece into the blade with the pattern riding against the duplicating jig. Adjust the position of the jig to control how close the blade cuts to the pattern. You then can sand right up to the line using a drum sander, or rout to the pattern using a flush-trimming bit.

Cut Circles Using this Handy Accessory

This system feature (**4–58**) takes the hassle out of cutting circles from 6" to 32" in diameter, thanks to its adjustable slider. To use the circle cutter, do the following:

1 Cut a square workpiece about ¹⁄₁₆" larger than the diameter of the circle you want to create. Carefully mark the exact center of the square, on the underside, by placing a straight-edge from corner to corner in each direction. Then drill a ¼"-diameter hole ¼" deep.

2 Loosen the setscrew that locks the slider in place, pull the slider to position the pivot pin, and retighten the setscrew. Now lower your workpiece onto the pivot pin so one edge is against the blade, and turn on the saw. Rotate the workpiece, as shown in **4–59**, to clip off the corners and create a perfect circle.

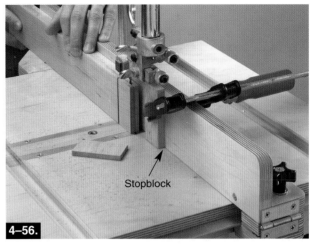

4–56.

Using the standard fence shown here, you can rip stock cleanly.

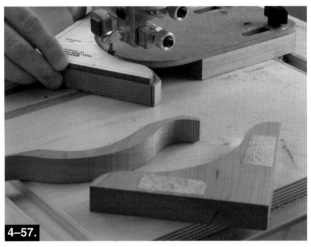

4–57.

Using the duplicating jig shown here, you can cut multiple curved pieces to the exact size and shape.

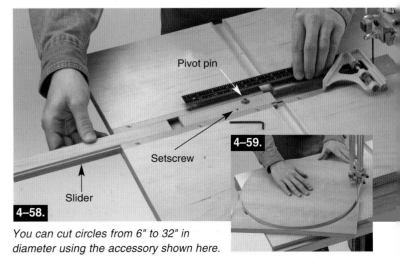

4–58.   **4–59.**

You can cut circles from 6" to 32" in diameter using the accessory shown here.

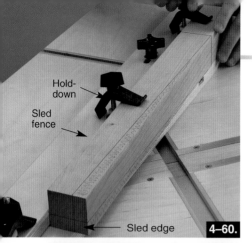

Hold-down

Sled fence

Sled edge

4–60.

Sled fence

Sled

4–61.

Guide bar

The jig shown here makes it easy to cut tapers with a bandsaw.

Taper with this Adjustable Jig

You might not think of using the bandsaw for cutting tapers, but this tapering jig (**4–60**) makes the process so effortless that you'll forget about tapering on the tablesaw. The jig holds your workpiece securely in place, and rides smoothly on a guide bar that slides in one of the table's T-track slots, as shown in **4–61**.

To precisely set the cutting angle, do the following:

1 Lay out the taper on your workpiece. Align those marks with the edge of the tapering jig's sled; then slide the fence against the workpiece, and tighten the two knobs that secure it to the sled. Secure the clamps to hold the workpiece in place.

2 With your leg blank in position, lower the jig's guide bar into the right-hand T-slot, well ahead of the blade. Then simply slide the jig forward to cut the waste away. Rotate your workpiece for the second taper, and cut again.

3 If you need to taper four sides of a part, such as a table leg, tape the cutoffs from your first two tapers back to your workpiece as shims. Now readjust the fence as necessary to align your layout marks for cuts three and four, and make the cuts. You'll also find this jig handy for putting a straight edge on rough-cut stock.

Crosscut Small Parts to Size

A bandsaw is the perfect tool for safely crosscutting small parts to size. The table system makes the process even easier by offering two slots— the aluminum T-tracks—for your miter gauge to slide in, as shown in **4–62**.

4–62.

Change Blades in No Time

It's plain to see that this bandsaw table offers lots of features you just can't get with the bare-bones stock saw. In addition to all of those great accessories, there's one important feature that's a little harder to see: You can change blades without removing the table! Just pop the blade insert out of the table and remove the slider. This exposes the slot in the stock bandsaw table, allowing you to easily slip blades in and out in normal fashion (**4–63**).

4–63.

CHAPTER 5

Drill-Press Options

T HE DRILL PRESS IS ONE OF *the most continually popular stationary power tools in the woodworking shop. Although it rarely captures the limelight, it's content to plug away, standing tall while performing workaday tasks. However unassuming, though, a drill press ranks only a step below a tablesaw in the hierarchy of valued workshop buddies.*

A drill press lets you precisely control drilling, often repeatedly punching one hole after another to an exact prescribed depth. And although most of us drill 90° to the surface, nearly all drill presses offer a tilting table to drill holes at many angles. But drilling holes is only a fraction of a drill press' capability. And that's what you'll find in this chapter—jigs and accessories that'll help you get more from this wonderful machine. Begin by checking out the angle-drilling jig that follows.

ADJUSTABLE ANGLE-DRILLING JIG

Make this simple tool (**5–1** and **5–2**) for your drill-press table and you'll never have to eyeball the angle of a hole again.

Build the jig as shown and dimensioned on **5–2**. The base (A) must be longer than your drill-press table, so the friction lid supports (one at each end

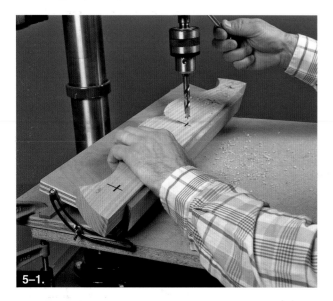

5–1.

of the jig) clear the table ends. Use a piano hinge to secure the adjustable support (B) to the plywood base (A). A pair of friction lid supports allow you to angle the support and lock it securely in position. The rest block (C) allows you to position the support parallel to the drill-press table.

To use the jig, do the following:

1 Loosen both wing nuts so the support can swivel freely.

2 Use a T-bevel or an adjustable triangle to set the required angle of the support to the drill bit.

3 Tighten the wing nuts to secure the support in place. Then clamp the jig to your drill-press table.

4 Drill a test hole to verify the angle. Once verified, drill the angled holes in your final workpiece.

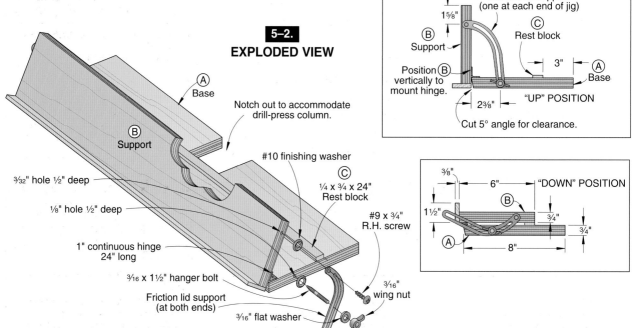

5–2.
EXPLODED VIEW

(A) Base

Notch out to accommodate drill-press column.

(B) Support

#10 finishing washer

(C) ¼ x ¾ x 24"
Rest block

³⁄₃₂" hole ½" deep

⅛" hole ½" deep

#9 x ¾"
R.H. screw

1" continuous hinge 24" long

³⁄₁₆ x 1½" hanger bolt

Friction lid support (at both ends)

³⁄₁₆" flat washer

³⁄₁₆"
wing nut

Friction lid support (one at each end of jig)

1⅝"

(B) Support

(C) Rest block

Position (B) vertically to mount hinge.

3"

(A) Base

"UP" POSITION

2⅜"

Cut 5° angle for clearance.

⅜"

6"

"DOWN" POSITION

(B)

1½"

¾"

¾"

(A)

8"

JIG THAT COVERS ALL THE ANGLES

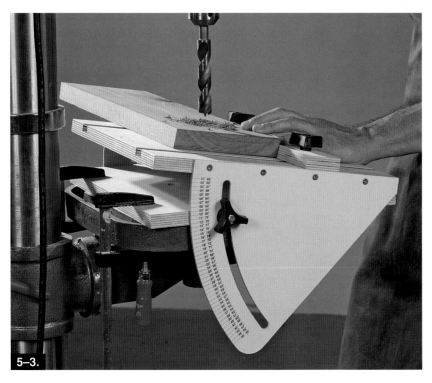

5–3.

E ven if you're fortunate
enough to have an accurate
drill press with a tilting table, the
one-way (side-to-side) tilt action
might not accommodate all the
angles or workpieces you deal
with. Worse still, the fixed tables
on some older drill presses give
you precision for 90° angles
and that's it.

This adjustable angle-drilling
jig (**5–3**) gives your drill press a
lot more versatility. It helps you
position and drill workpieces
consistently—the essential job
of any drill press—and even can
tackle such difficult tasks as
compound angles.

Two Baltic birch panels—one
base and one platform—make up

most of the jig. The base clamps
or bolts to the drill press table,
and the platform pivots to angles
between 0° and 45°.

You'll find the angle scale
on *page 127*. Simply cut it out,
enlarge it to 200%, and use spray
adhesive to glue it to a piece
of ⅛" plywood or hardboard.
Set another ⅛" panel (for the
righthand angle support)
underneath the first and cut them
out together, oriented back to
back. Drill starter holes at the
ends of the curved slot; then
cut the center portion out with a
scrollsaw or jigsaw.

Materials List for Jig That Covers All the Angles

PART	FINISHED SIZE			MTL.	QTY.
	T	W	L		
A base	¾"	14"	14"	BP	1
B platform	¾"	14"	14"	BP	1
C angle supports	⅛"	9¾"	11⅞"	HB	2
D fence	¾"	2"	14"	BP	1

Materials Key: BP = Baltic birch plywood;
HB = hardboard.
Supplies: ⅜" x 3/16" mini channel (2 pcs.
14" long); #6–32 T-nuts and mating #6–32
x ¼" flathead machine screws (10 each);
⅜" x 1½" square-head bolts (2); ⅜" x 3"
carriage bolts (2); ⅜" three-wing plastic
knobs (6); ⅜" flat washers (8); ⅜ x 2"
threaded rod (2); #8 x 1" flathead wood
screws (8); 11/16" x 14" continuous hinge
with mounting screws.

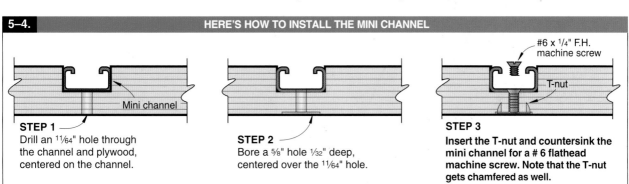

5–4. | **HERE'S HOW TO INSTALL THE MINI CHANNEL**

STEP 1
Drill an 11/64" hole through
the channel and plywood,
centered on the channel.

Mini channel

STEP 2
Bore a ⅝" hole 1/32" deep,
centered over the 11/64" hole.

#6 x ¼" F.H.
machine screw

T-nut

STEP 3
**Insert the T-nut and countersink the
mini channel for a # 6 flathead
machine screw. Note that the T-nut
gets chamfered as well.**

Mounting the table correctly requires the use of a simple indexing template, shown at a reduced scale in **5–7**. Cut a notch at one corner (equal to half the column diameter), so you can line up the template edge with the centerline on the jig.

5–5.

EXPLODED VIEW

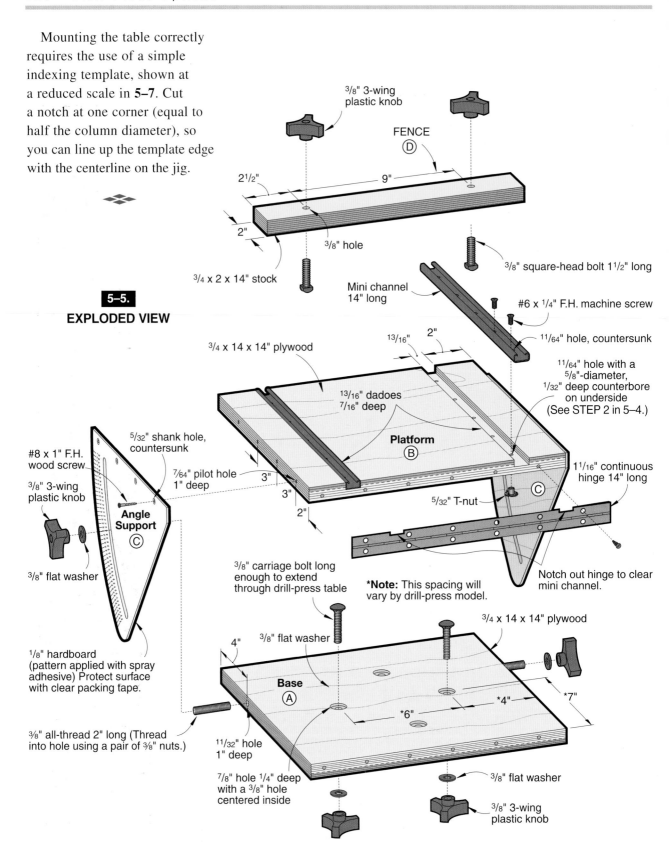

3/8" 3-wing plastic knob

FENCE
(D)

2¹/₂"

9"

2"

3/8" hole

3/4 x 2 x 14" stock

3/8" square-head bolt 1¹/₂" long

Mini channel 14" long

#6 x 1/4" F.H. machine screw

11/64" hole, countersunk

13/16"   2"

3/4 x 14 x 14" plywood

13/16" dadoes 7/16" deep

11/64" hole with a 5/8"-diameter, 1/32" deep counterbore on underside (See STEP 2 in 5–4.)

Platform
(B)

1¹/₁₆" continuous hinge 14" long

5/32" shank hole, countersunk

#8 x 1" F.H. wood screw

3/8" 3-wing plastic knob

7/64" pilot hole 1" deep

3"

3"

5/32" T-nut

Angle Support
(C)

2"

(C)

Notch out hinge to clear mini channel.

3/8" flat washer

1/8" hardboard (pattern applied with spray adhesive) Protect surface with clear packing tape.

3/8" carriage bolt long enough to extend through drill-press table

*Note: This spacing will vary by drill-press model.

3/4 x 14 x 14" plywood

4"

3/8" flat washer

Base
(A)

*4"

*7"

*6"

3/8" all-thread 2" long (Thread into hole using a pair of 3/8" nuts.)

11/32" hole 1" deep

7/8" hole 1/4" deep with a 3/8" hole centered inside

3/8" flat washer

3/8" 3-wing plastic knob

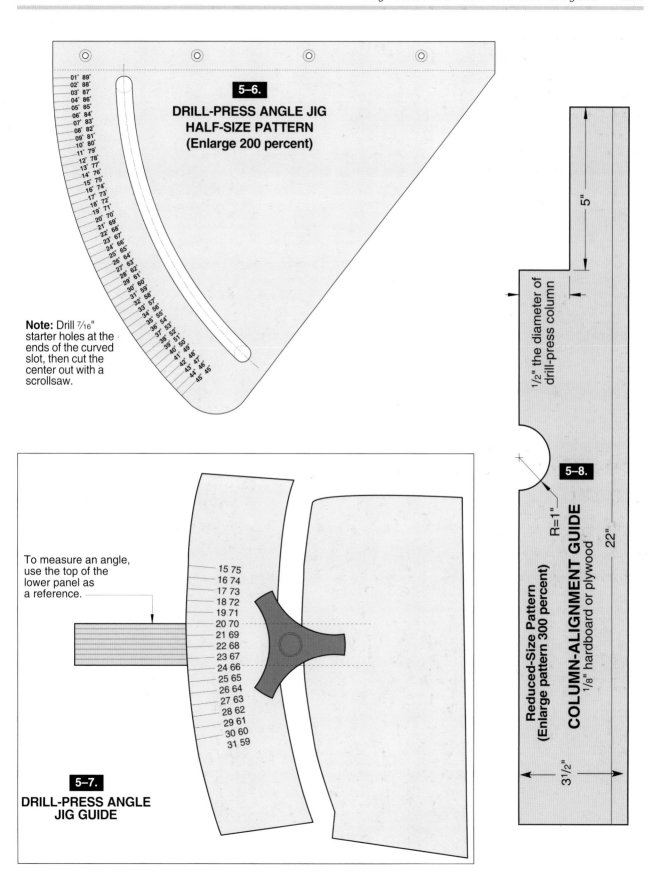

5–6.

**DRILL-PRESS ANGLE JIG
HALF-SIZE PATTERN
(Enlarge 200 percent)**

Note: Drill 7/16" starter holes at the ends of the curved slot, then cut the center out with a scrollsaw.

To measure an angle, use the top of the lower panel as a reference.

5–7.

**DRILL-PRESS ANGLE
JIG GUIDE**

5–8.

1/2" the diameter of drill-press column

R=1"

Reduced-Size Pattern
(Enlarge pattern 300 percent)

COLUMN-ALIGNMENT GUIDE
1/8" hardboard or plywood

5"

22"

3 1/2"

DRUM-SANDING TABLE

This handy table (**5–9** and **5–10**) helps you in three important ways. First, it has a dust-collection port for capturing fine dust before it becomes airborne. The table also accepts inserts that you size to tightly fit your sanding drums. That provides workpiece support and improves dust-collection efficiency. And, with storage areas for sanding drums and inserts, you'll always be organized.

To make the opening in the top for the inserts, first cut a 3½"-square hole using a jigsaw. Then, use your router to form the ¼" rabbet around the opening. Square the corners with a chisel, or leave them round and sand the corners of the inserts to fit the opening. We made our inserts from ¼" melamine-coated polyboard, but any hardboard or plywood will do.

To use the table, install the applicable hardboard insert and position the table that the

5–9.

sanding drum is centered with the hole in the insert. Clamp the table in place, and adjust the drill-press table, if necessary, to square the sanding surface with

the drill-press spindle. Attach the hose from your vacuum, and you're set to go.

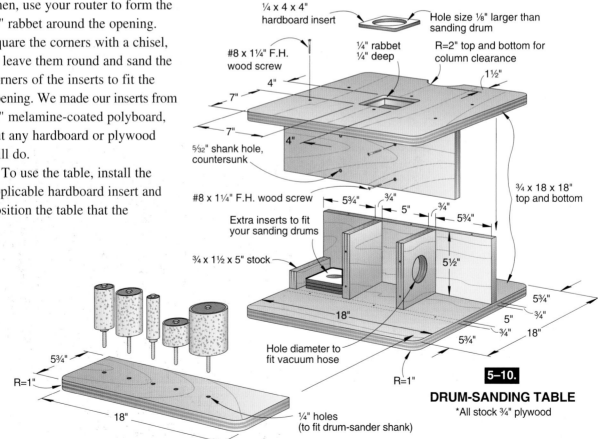

¼ x 4 x 4"
hardboard insert

Hole size ⅛" larger than sanding drum

#8 x 1¼" F.H. wood screw

¼" rabbet ¼" deep

R=2" top and bottom for column clearance

1½"

7"

4"

7"

4"

⁵⁄₃₂" shank hole, countersunk

¾ x 18 x 18" top and bottom

#8 x 1¼" F.H. wood screw

5¾"   ¾"
5"   ¾"   5¾"

Extra inserts to fit your sanding drums

¾ x 1½ x 5" stock

5½"

5¾"
5"   ¾"
5¾"   ¾"   18"

18"

Hole diameter to fit vacuum hose

R=1"

5–10.

DRUM-SANDING TABLE
*All stock ¾" plywood

5¾"

R=1"

18"

¼" holes
(to fit drum-sander shank)

SUPER-VERSATILE DRILL-PRESS TABLE

Quality workmanship depends in large part on the accuracy of your tools. With this in mind, we set out to make your already hard-working drill press into the ultimate precision-machining center (**5–11** and **5–12**). As an option, see *page 138* for another multi-function drill-press table.

This project became a reality after *WOOD®* magazine's staff designed and tested several prototypes before settling on this L-shaped table coupled with a pair of firm fences and hold-down clamps. As shown on *pages 133* and *134*, this setup allows you to perform numerous machining processes with impressive precision.

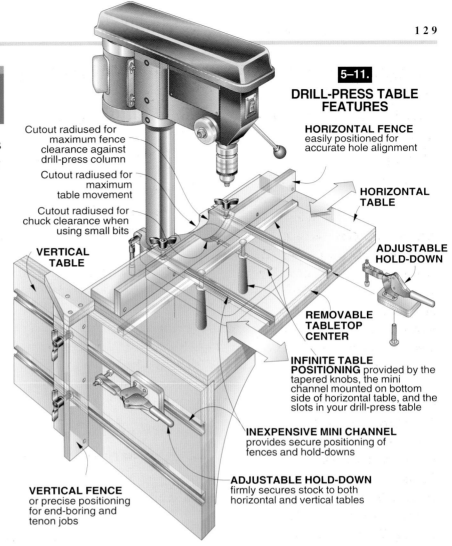

Cutout radiused for maximum fence clearance against drill-press column

Cutout radiused for maximum table movement

Cutout radiused for chuck clearance when using small bits

VERTICAL TABLE

5–11.

DRILL-PRESS TABLE FEATURES

HORIZONTAL FENCE easily positioned for accurate hole alignment

HORIZONTAL TABLE

ADJUSTABLE HOLD-DOWN

REMOVABLE TABLETOP CENTER

INFINITE TABLE POSITIONING provided by the tapered knobs, the mini channel mounted on bottom side of horizontal table, and the slots in your drill-press table

INEXPENSIVE MINI CHANNEL provides secure positioning of fences and hold-downs

ADJUSTABLE HOLD-DOWN firmly secures stock to both horizontal and vertical tables

VERTICAL FENCE or precise positioning for end-boring and tenon jobs

Start with the Horizontal and Vertical Table Pieces

1 Cut the horizontal table (A) and vertical table pieces (B) to the sizes listed in the Materials List. (Due to its stability and strength, we used ¾" Baltic birch plywood.)

2 To make the radiused support (C), cut a piece of ¾"-thick plywood to 5¼ × 10¼". Mark a 9" radius and cut the support to the shape shown on **5–14,** on *page 130*.

3 Cut the horizontal table tops (D, E) to size. (After you have used the drill-press table awhile, the center piece will have several holes drilled into it. For ease in replacing the tabletop center [E], we constructed the tabletop out of three pieces rather than one.)

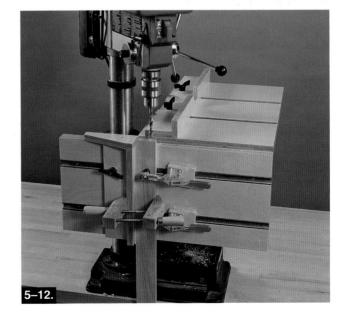

5–12.

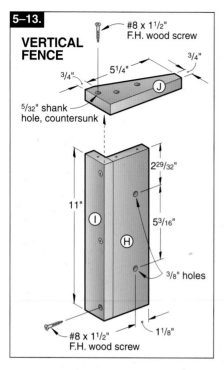

5–13.

VERTICAL FENCE

#8 x 1¹/₂" F.H. wood screw

3/4"   3/4"

5¹/₄"

J

⁵/₃₂" shank hole, countersunk

2²⁹/₃₂"

11"   5³/₁₆"

I   H   ³/₈" holes

#8 x 1¹/₂" F.H. wood screw   1¹/₈"

4 Using a hacksaw, crosscut five pieces of metal mini channel to the lengths listed on **5–16**. Drill *countersunk* mounting holes at 4" intervals through the mini channel for screwing to the plywood tables later.

5 Mark the location of the groove on the bottom surface of the horizontal table (A). Position the table on the top of your metal drill-press table. Looking through the slots on your drill-press table, verify that the marked groove crosses at least two of the grooves in your metal table. If not, adjust the position of the groove.

6 Using a dado blade in your tablesaw, cut ¹³/₁₆" dadoes, rabbets, and grooves (just wide enough to house the mini channel cut earlier) in parts A, B, and D, where dimensioned on **5–14** and **5–16**.

7 Drill countersunk holes, and glue and screw the horizontal table (A) to the *inside* of the vertical table (B) in the configuration shown on the drawings. Drill more countersunk mounting holes, and drive screws through the *outside* surface of A and B to secure the radiused support (C) to keep parts A and B exactly perpendicular to each other.

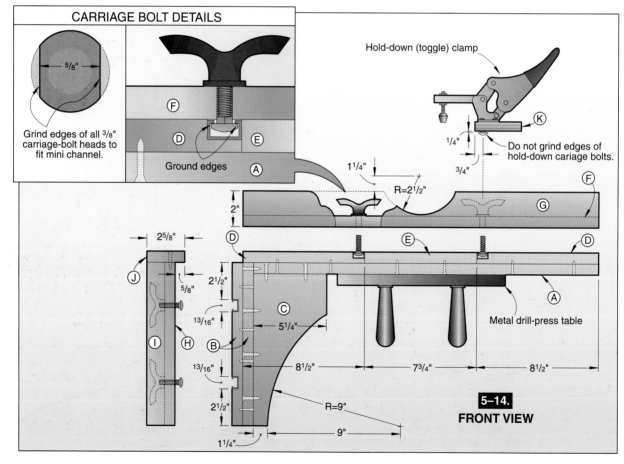

CARRIAGE BOLT DETAILS

⁵/₈"

Grind edges of all ³/₈" carriage-bolt heads to fit mini channel.

F

D   E

Ground edges   A

Hold-down (toggle) clamp

¹/₄"   K

³/₄"   Do not grind edges of hold-down cariage bolts.

1¹/₄"

R=2¹/₂"   F

2"   G

D   E   D

2⁵/₈"

J   2¹/₂"

⁵/₈"

13/16"   C

I   H   B   5¹/₄"   A

Metal drill-press table

13/16"   8¹/₂"   7³/₄"   8¹/₂"

2¹/₂"

R=9"   **5–14.**

1¹/₄"   9"   **FRONT VIEW**

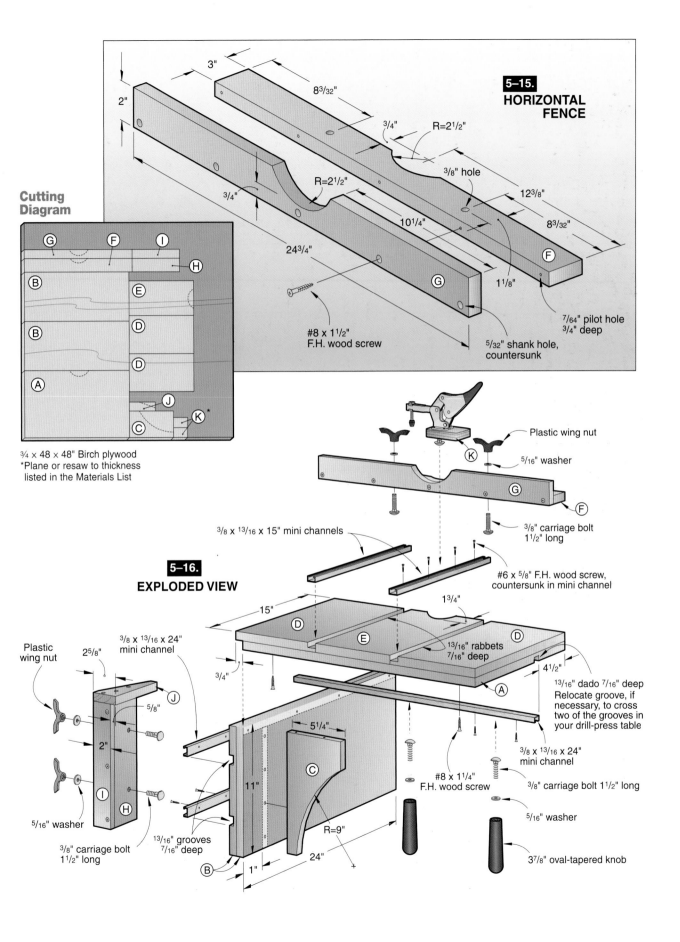

Cutting Diagram

G F I
H
B
E
B
D
D
A
J
C K *

¾ × 48 × 48" Birch plywood
*Plane or resaw to thickness
listed in the Materials List

5–15.
HORIZONTAL FENCE

3"
8³/₃₂"
2"
¾"
R=2¹/₂"
3/8" hole
12³/₈"
8³/₃₂"
¾"
R=2¹/₂"
10¹/₄"
24³/₄"
F
1¹/₈"
G
7/64" pilot hole
3/4" deep
#8 x 1¹/₂"
F.H. wood screw
5/32" shank hole,
countersunk

Plastic wing nut
K
5/16" washer
G
F
3/8" carriage bolt
1¹/₂" long

³/₈ x ¹³/₁₆ x 15" mini channels
#6 x 5/8" F.H. wood screw,
countersunk in mini channel

5–16.
EXPLODED VIEW

15"
1³/₄"
D
E
D
13/16" rabbets
7/16" deep
4¹/₂"
³/₈ x ¹³/₁₆ x 24"
mini channel
2⁵/₈"
¾"
13/16" dado 7/16" deep
Relocate groove, if
necessary, to cross
two of the grooves in
your drill-press table
Plastic
wing nut
J
5/8"
2"
A
³/₈ x ¹³/₁₆ x 24"
mini channel
3/8" carriage bolt 1¹/₂" long
I
H
5/16" washer
5/16" washer
3/8" carriage bolt
1¹/₂" long
13/16" grooves
7/16" deep
5¹/₄"
C
11"
B
1"
24"
R=9"
#8 x 1¹/₄"
F.H. wood screw
3⁷/₈" oval-tapered knob

8 Drill countersunk holes through the *inside* surfaces of the horizontal (A) and vertical (B) tables to secure the outside piece of the vertical table and tabletops D/E. The tabletop sides (D) are glued and screwed to the horizontal table (A), and the removable tabletop center (E) is just screwed in place. Glue and screw the outside vertical table (B) to the assembly. Cut a 2½" radius centered on the back edge of A/E, as shown on **5–16**.

9 Screw the five pieces of mini channel in place.

Build a Pair of Fences for Precision Alignment

1 Cut the horizontal fence pieces (F, G) to size. Mark a 2½" radius on each, where shown on **5–15**, and bandsaw and drum-sand the radii to shape. The radius centered on the horizontal table fence (G) allows the drill-press chuck to come closer to the work-piece, which is especially helpful when using a small bit in the chuck. The radius in the fence base (F) allows you to move the fence closer to the drill-press column when drilling into wide stock.

2 Mark a pair of hole center-points on the fence base (F). Before drilling the holes, verify that the marked points are centered over the dadoed openings in the table top. Improperly located, the carriage bolts will bind in the mini channel. Adjust if necessary, and drill the ⅜" holes.

3 Drill countersunk mounting holes, and screw part G to F, checking for square.

4 Repeat the process to construct the vertical fence (H, I). Cut the end guide (J) to shape, drill the mounting holes, and screw it to the top end of the vertical fence.

5 To mount the fences to the tables, start by grinding two opposing edges of four ⅜" carriage bolts until the bolt heads slide easily in the metal mini channel. See the Carriage Bolt details in **5–14** for reference. Then, insert the bolts through the fence pieces (F, H), through flat wash-ers, and into the plastic wing nuts (knobs). The fence assemblies should slide back-and-forth in the tables easily, but without slop.

Materials List for Super-Versatile Drill-Press Table

PART	FINISHED SIZE			MTL.	QTY.
	T	W	L		
TABLE ASSEMBLIES					
A horizontal table	¾"	15"	24"	BP	1
B vertical table pieces	¾"	11"	24"	BP	2
C support	¾"	5¼"	10¼"	BP	1
D tabletop sides	¾"	8½"	15"	BP	2
E tabletop center	¾"	7¾"	15"	BP	1
FENCES AND HOLD-DOWNS					
F horizontal fence base	¾"	3"	24¾"	BP	1
G horizontal fence	¾"	2"	24¾"	BP	1
H vertical fence base	¾"	3"	11"	BP	1
I vertical fence	¾"	2"	11"	BP	1
J end guide	¾"	2⅝"	6"	BP	1
K hold-down bases	⅝"	2¼"	3¼"	BP	2

Material Key: BP = Baltic birch plywood
Supplies: #6 x ⅝", #8 x 1¼", #8 x 1 1/2" flathead wood screws; ⅜" x 1½" (6) and 5⁄16" x 1" (2) carriage bolts; two 2"-reach hold-down (toggle) clamps; four plastic wing nuts (knobs); two 3⅛"-long, oval-tapered knobs; 5⁄16" locknuts (2); 5⁄16" flat washers (6); ⅜ x 13⁄16 x 24" mini channel (3); ⅜ x 13⁄16 x 15" mini channel (2); #12 x ⅝" panhead sheet metal screws (8); clear finish.

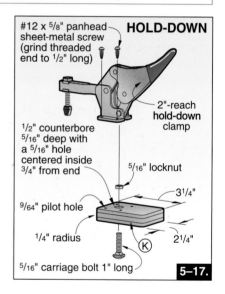

HOLD-DOWN
#12 x ⅝" panhead sheet-metal screw (grind threaded end to 1/2" long)
2"-reach hold-down clamp
½" counterbore 5⁄16" deep with a 5⁄16" hole centered inside ¾" from end
5⁄16" locknut
3¼"
9⁄64" pilot hole
¼" radius
2¼"
K
5⁄16" carriage bolt 1" long

5–17.

The Hold-Downs Come Next

1 Cut the clamp hold-down bases (K) to size from ⅝" stock. (We resawed some of the ¾" plywood we used for the other pieces.) (See **5–17**.) For holding ¾"-thick stock later, the bases must be less than ¾" thick. Cut or sand a ¼" radius on each corner.

2 Drill a ½" hole ⁵⁄₁₆" deep in the top of each hold-down base. Now, drill a ⁵⁄₁₆" hole, centered through the middle of the counterbore.

3 Insert a ⁵⁄₁₆" carriage bolt into the bottom surface of each hold-down base. (For ease in using the hold-down clamps, we found that the hold-down bases must swivel on the tops of the tables. In order to do this, we had to use a ⁵⁄₁₆" carriage bolt instead of the ⅜" bolts used on the fences.)

4 Grind the threaded end of eight #12 x ⅝" pan-head sheet-metal screws until the threaded end is ½" long. Then, drill pilot holes, and secure one hold-down clamp to each base (K) with four of the screws.

Secure the Assembly to Your Drill Press

1 Sand all the parts. Mask off the metal mini channel and apply a clear finish to the wood parts. (We used an aerosol lacquer.)

2 Secure the wooden drill-press table to the metal table with a pair of ⅜" carriage bolts (with the edges of the head ground as previously explained and shown on the Carriage Bolt details in **5–14**), flat washers, and tapered knobs. (We found the long tapered knobs easy to grab and the table easier to relocate when needed.) Slide the fences with knobs attached onto their mating mini channel.

Put Your New Table To Work

Illus. **5–18** to **5–22** show several uses for our multi-purpose drill-press table. Once you've built your own, you may discover other applications.

5–18.

Making round-cornered mortises and the mating tenons is easy to accomplish using our fence system. As shown here, drill a hole with a Forstner bit to form the mortise.

5–19.

Right: To form a tenon to mate with the round-cornered mortise formed at left, use a tablesaw and a V-grooved support to cut the shoulders. Then, as shown here, use the hold-downs and clamp to secure the stock to the vertical table and fence. Use a plug cutter to cut a round tenon on the end of the stock.

Continued on following page.

Left: *For drilling holes in the ends of stock, switch to the vertical fence with hold-downs. As shown here, we drilled a pair of dowel holes in the ends of a rail to mate with the stile drilled in* **5–22**. *To do this, mark the dowel-hole centerpoints on the ends of the rail. Next, use the hold-downs to clamp the rail to the vertical table. Loosen the collar connecting the drill-press table to the drill-press column. Swing the drill-press table to the side until the drill bit aligns directly over the marked centerpoint. (You'll also have to move the rail to get it under the bit.)*

If the bit won't align, rotate the drill-press head on the column for proper alignment as we had to do for this photo. Adjust the position of the fence until the bit aligns with the marked centerpoint. Check that the fence is plumb. Use the hold-downs to secure the rail to the vertical table. Use a handscrew clamp to secure the piece to the fence. Without the handscrew clamp, the stock being drilled can slip slightly down the vertical table. Finally, drill the holes.

5–20.

Below: *You can use the setup shown in* **5–21** *to drill holes in the edge of a piece of stock, as shown here. This setup is especially useful for drilling dowel holes in the edge of stiles.*

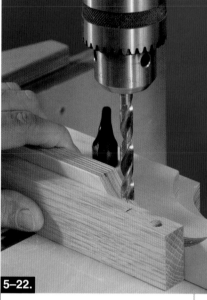

5–22.

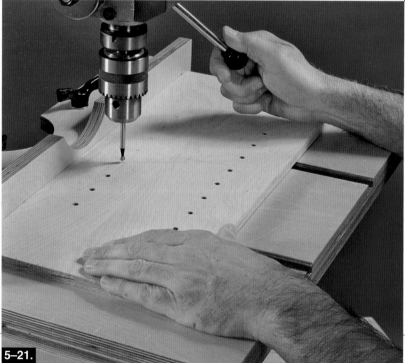

5–21.

Left: *Drilling numerous holes all in a straight line is easy using the horizontal table and fence. For instance, if you need holes exactly 2" from the edge of a piece of stock, position the inside edge of the fence exactly 2" from the center of the bit. Now, just move the stock along the fence and all the holes will be drilled exactly 2" from the edge of the stock, as shown in here.*

BALL-DRILLING JIG

5–23.

Some projects (although few and far between) call for drilling perfectly centered holes in hardwood balls. While you can hold a ball securely in a wood handscrew, centering it under the bit on your drill press is difficult.

Solve both problems with this quick-to-build jig (**5–23** and **5–24**). To make it easy to clamp the jig to your drill press, make the length of the base the same as the width of your drill-press table. Do the following:

JIG HOLE GUIDE	
Ball dia.	Hole Dia.
3/4"	1/2"
1"	3/4"
1 1/4", 1 1/2"	1"
1 3/4", 2"	1 1/2"
2 1/4", 2 1/2"	1 3/4"
2 3/4", 3"	2"

1 Drill the counterbored holes for the carriage bolts and the large holes in the base and cap. (See the Jig Hole Guide in **5–24**.)

2 Insert the bolts and center the jig by lowering the bit back into the hole, and holding it there while you clamp the jig to your drill-press table.

3 Place the ball over the hole, slide the cap down on the protruding carriage bolts, and tighten it onto the ball with washers and wing nuts. Chuck in the proper bit and drill away.

JIGS FOR DOWELS

Drilling holes into the sides and ends of dowels can challenge even the best woodworkers. The problem: trying to hold round stock in position. Try these simple jigs (**5–25** and **5–26**) to hold the dowels rock-steady. To make them, do the following:

1 Both jigs rely on a V-groove that you cut into a piece of 3/4 x 3 x 20" scrap wood. To cut the V, tilt the blade of your table-saw to 30° from vertical, and set the cutting depth to 3/8", as shown in Cutting the V-Groove detail in **5–27**. Adjust the fence to place the top of the cut on the center-line of the board. Make one pass over the saw blade, turn the board around, and run it through again, creating a 60° V-groove.

5–24.
DRILLING JIG

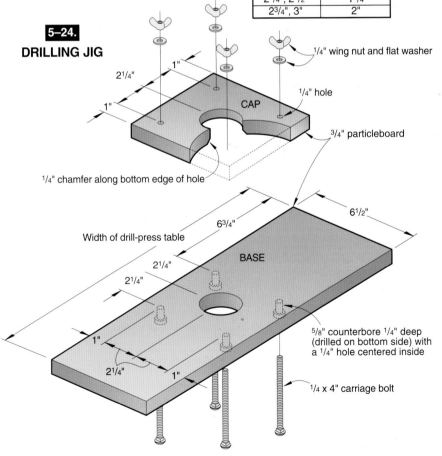

2 1/4"   1"

1"

CAP

1/4" wing nut and flat washer

1/4" hole

3/4" particleboard

1/4" chamfer along bottom edge of hole

6 1/2"

6 3/4"

Width of drill-press table

2 1/4"

BASE

2 1/4"

1"

5/8" counterbore 1/4" deep (drilled on bottom side) with a 1/4" hole centered inside

2 1/4"   1"

1/4 x 4" carriage bolt

5–25.

Clamped to the drill-press table, this grooved piece of scrapwood holds dowels steady for horizontal drilling.

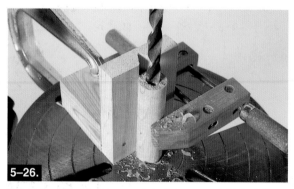

5–26.

Screw the shorter V-groove block to an upright to drill straight down into the end of the dowel.

5–27. **VERTICAL JIG**

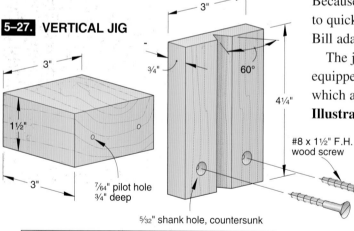

3"

3"

1½"

3"

¾"

60°

4¼"

⁷⁄₆₄" pilot hole ¾" deep

⁵⁄₃₂" shank hole, countersunk

#8 x 1½" F.H. wood screw

CUTTING THE V-GROOVE DETAIL

Fence

1½"

30°

⅜"

Tablesaw

2 Crosscut a 4¼" piece off one end. The longer piece holds your dowels flat on the table, as shown in **5–25**. The shorter piece forms the basis for the second jig, which holds dowels vertically for drilling into their ends (**5–26**).

3 To complete the vertical jig, cut a 1½ x 3 x 3" block, clamp it to the back of the 4¼" piece, and check to make sure that the V-groove is perpendicular to the base. Then glue and screw the two pieces together, as shown in **5–27**.

STEP-AND-REPEAT JIG

Having worked for more than 40 years as a printer, *WOOD®* magazine reader Bill Lacey is familiar with the process called "step and repeat." In printing, small images, such as business cards, are reproduced in rows and columns to fill a full sheet of paper. Because he builds toys in small batches and needs to quickly and accurately repeat drilling operations, Bill adapted this idea to his woodworking.

The jig (**5–28**) has two components: the carriage, equipped with rear and end fences, and the spacers, which allow you to position your workpiece. **Illustrations 5–29 to 5–32** show how to use the jig.

Build the carriage as shown in **5–28**. The size of the carriage can vary according to the size of pieces you are drilling and the maximum reach of your drill press. You can cut spacers to standard widths ahead of time, or custom-make spacers for each different job. Bill took the first approach, cutting spacers in widths from ⅛" to 1" in ¹⁄₁₆" increments, and from 1" to 6" in 1" increments. Cut all the spacers of the same width at the same time to ensure uniformity.

Illustrations 5–29 to 5–32 demonstrate using the jig to drill holes in a cribbage board. We positioned the workpiece by inserting spacers between the fences and

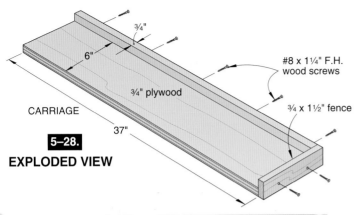

5–28.
EXPLODED VIEW

CARRIAGE

¾"
6"
¾" plywood
37"

#8 x 1¼" F.H. wood screws

¾ x 1½" fence

END-BORING JIG

Sure, it's easy to drill into the faces or edges of most rectangular stock or dowels. But what about boring into the ends of long rectangular stock or dowels? This simple jig (**5–33** and **5–34**) does the trick, with perfect precision and control.

To accommodate long stock with a benchtop drill press, mount the drill press at the end of your bench. Then, rotate its head to clear its worktable and the workbench top.

5–29.

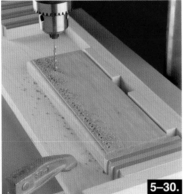

5–30.

Locate the first marked hole under the bit, and clamp the carriage in place. Drill the first row of holes.

Spacers inserted between the rear fence and the workpiece position it for drilling the second row of holes.

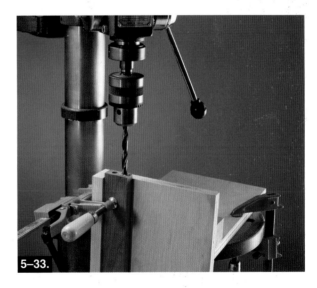

5–33.

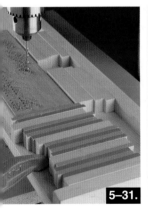

5–31.

5–32.

Insert more spacers between the rear fence and the workpiece to position it for the third row of holes.

One more row of spacers behind the workpiece positions it for the final row.

the workpiece to progressively move it away from the end fence and out from the rear fence. The green spacers move the piece in ⅛" increments, the yellow spacers in ½" increments.

1"
2"
¾"
#8 x 1¼" F.H. wood screw
12"
12"
1½"
1¼' #8 x 2" F.H. wood screws
12"
8"
¾"
8"

5–34.
END-BORING JIG

A FEATURE-PACKED TABLE

Although indispensable in a woodworking shop, most drill presses come equipped with a table more suited to metal-working. This add-on table with fence (5–35 to 5–37) sets things straight, supplying everything the cast-iron table on your drill press lacks.

Start with the Table

1 For the base (A), cut two 14½ x 29½" pieces of ½" plywood. (We used Baltic birch plywood for its flatness and absence of voids. You also can use regular birch plywood or medium-density fiberboard.) Glue and clamp the pieces together, keeping their ends and edges flush, where shown on **5–38**.

2 From ¼" tempered hardboard, cut the top sides (B), top front (C), and top back (D) to the sizes listed in the Materials List on *page 144*. Mark the ⅜"-radius finger notch on the front edge of part D, where shown on **5–38**. Cut and sand it to shape. (The notch makes it easy to remove the insert [E].) Now spread glue on the backs of the hardboard top parts, and glue and clamp them to the 1"-thick plywood base (**5–39**).

3 Draw the 3¼"-radius cutout at the rear of the table, where shown on **5–38**. Bandsaw or jigsaw and sand it to shape.

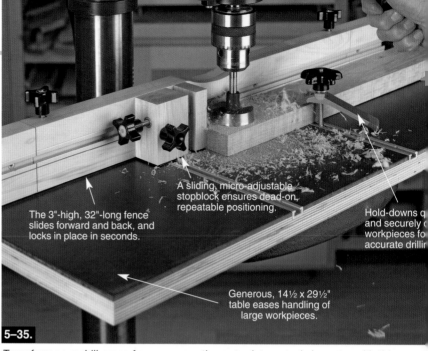

The 3"-high, 32"-long fence slides forward and back, and locks in place in seconds.

A sliding, micro-adjustable stopblock ensures dead-on, repeatable positioning.

Hold-downs q and securely c workpieces fo accurate drillir

Generous, 14½ x 29½" table eases handling of large workpieces.

5–35.

Transform your drill press from a supporting actor into a workshop star with this multi-function accessory.

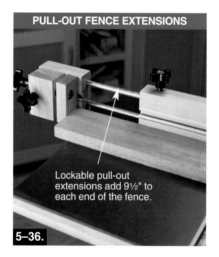

PULL-OUT FENCE EXTENSIONS

Lockable pull-out extensions add 9½" to each end of the fence.

5–36.

REPLACEABLE TABLE INSERT

A replaceable insert provides full workpiece support when drilling or sanding.

5–37.

4 To locate the 3½ x 3½" cutout in the add-on table base, install a ⅛" bit in the drill-press chuck, center your metal drill-press table under the bit, and lock the table in place. Position the add-on table to center the bit in the recess for the insert (E) created by parts B, C, and D. If the metal drill-press table protrudes beyond the front edge of the add-on table, slide the add-on table forward until the two are flush.

Clamp the add-on table in place. Now drill a ⅛" hole all the way through the base (A). Remove the add-on table, and turn it upside down. Mark the 3½ x 3½" cutout centered on the ⅛" hole. Then drill ⅜" holes at the corners, and cut the opening with a jigsaw. Now cut the insert (E) to the size listed in the Materials List.

5 For a drill press with slots through its metal table, cut the ¾" x ⅜" groove for the mini

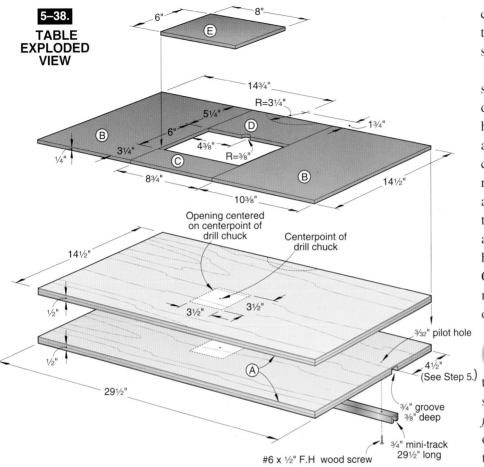

5–38.
TABLE EXPLODED VIEW

8"
6"
E

14¾"
R=3¼"
5¼"
1¾"
6"
D
B
4⅜"
R=⅜"
¼"
3¼"
C
8¾"
B
14½"
10⅜"

Opening centered on centerpoint of drill chuck
Centerpoint of drill chuck

14½"
½"
3½"
3½"
½"

29½"
A

³⁄₃₂" pilot hole

4½"
(See Step 5.)

¾" groove ⅜" deep

#6 x ½" F.H wood screw

¾" mini-track 29½" long

GLUING UP THE TABLE

Caul
D
B
B
A
C
Masking Tape
Bearer
Caul

5–39.

With glue applied to their bottom surfaces, position top parts B, C, and D on the 1"-thick base (A). To keep them from shifting, tape the top parts to each other and to the base with masking tape. Then clamp the top and base between ¾-thick cauls and 2 x 3 bearers.

channel in the bottom of the add-on table, where shown on **5–38**.

For a drill press without slots in its metal table, drill two ¼" mounting holes. Locate the holes about halfway between the center of the table and its rear edge and as far apart as possible. Then clamp the add-on table in place as before, and trace the hole locations on its bottom. Cut the groove for the mini channel so it passes over the holes.

6 Turn the table over, and cut dadoes for the top mini channels, where shown on **5–41**, on the *following page*. (The dadoes are centered on the joint lines between parts B and parts C and D.)

Now Make the Fence

1 Cut the base blank (F), face blank (G), lower rear blank (H), and upper rear blank (I) to size. Install a ⅜" dado blade in your tablesaw and position the fence to cut centered grooves in the thickness of parts H and I, where shown on **5–44** and **5–45**, on *pages 141 and 142*. Then cut ³⁄₁₆"-deep grooves in these parts, and mark the face that was against the saw fence.

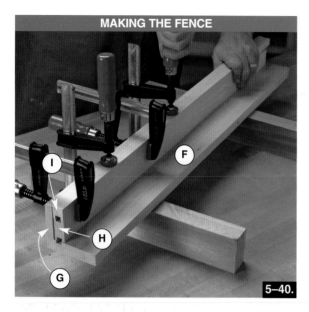

MAKING THE FENCE

5–40.

With their marked faces against the back of the face blank (G), glue the lower rear blank (H) and the upper rear blank (I) to each other and to the base blank (F) and face blank (G). Apply clamping pressure in two directions.

When cutting the top and bottom grooves in the lower rear blank, keep the marked face against the fence for both cuts. Now without changing the setup, cut a mating groove in the base blank.

2 Glue and clamp the face blank (G) to the base blank (F), where shown on **5–45**, on *page 142*. Make sure the face blank is square to the base blank. With the glue dry, glue and clamp the lower rear blank (H) and upper rear blank (I) in place, as shown in **5–40**. Before the glue dries, run a length of ⅜" steel rod in and out of the square holes to clear away any excess glue.

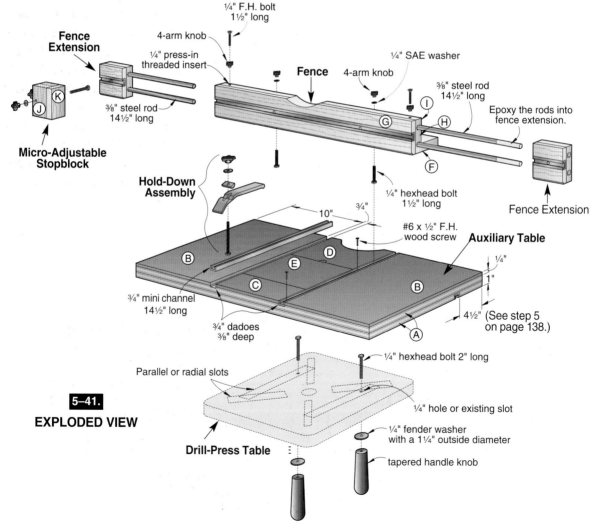

¼" F.H. bolt 1½" long

Fence Extension

4-arm knob

¼" press-in threaded insert

⅜" steel rod 14½" long

¼" SAE washer

Fence

4-arm knob

⅜" steel rod 14½" long

Epoxy the rods into fence extension.

Micro-Adjustable Stopblock

Hold-Down Assembly

Fence Extension

10"

¾"

¼" hexhead bolt 1½" long

#6 x ½" F.H. wood screw

Auxiliary Table

¼"

1"

¾" mini channel 14½" long

¾" dadoes ⅜" deep

4½" (See step 5 on page 138.)

¼" hexhead bolt 2" long

Parallel or radial slots

5–41.
EXPLODED VIEW

Drill-Press Table

¼" hole or existing slot

¼" fender washer with a 1¼" outside diameter

tapered handle knob

Tips on Using Threaded Inserts

Shop fixtures and jigs often require the installation of various clamping or adjustment knobs. That's when you'll reach for threaded inserts. Commonly available in sizes from #8–32 (a #8 screw body with 32 threads per inch) to ⅜"–16 (a ⅜" screw body with 16 threads per inch), there are two basic types: thread-in and press-in, shown in 5–42.

Use thread-in inserts in softer woods and plywood where their coarse outside threads cut easily into the surrounding wood. Simply drill a hole sized for the body of the insert, and screw it into place. In very hard woods, such as white oak and maple, or when the insert is close to the edge of a part and screwing it in may split the wood, drill a hole slightly larger than the outside thread diameter and epoxy the insert in place. To protect the internal threads from epoxy, cover the end of the insert, as shown in 5–43.

Press-in inserts, with their barbed exteriors, work well in hardwoods, softwoods, and plywood. Drill a hole sized for the body of the insert and press it into place with a clamp or tap it in with a hammer and a block of wood. For applications in which the clamping action tends to push the insert out of the wood, such as the knobs that tighten down on the drill-press fence extension rods, drill a hole that engages just the tips of the insert barbs and epoxy it in place.

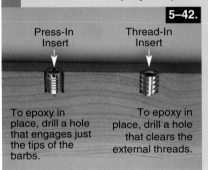

5–42.

Press-In Insert   Thread-In Insert

To epoxy in place, drill a hole that engages just the tips of the barbs.

To epoxy in place, drill a hole that clears the external threads.

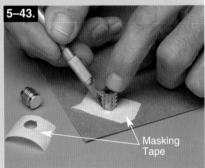

5–43.

Masking Tape

3 Cut a ¾" groove ⅜" deep for the mini channel in the fence face (G), where shown on **5–45**. Then cut a ⅛ × ⅛" sawdust-relief rabbet along the bottom edge of the fence face.

4 Trim one end of the assembled fence blank square, and then cut it into three pieces, where shown on **5–44**, making a 22½"-long fence and two 3½"-long extensions. Now cut off the base portions of the extensions, where shown on **5–45**.

5 Bending a fairing strip to join their endpoints and center-points, mark the centered radius cutouts on the top of the fence and the back of its base (F), where shown on **5–44**. Bandsaw or jigsaw them, and sand them to shape. Then drill ¼" holes for the bolts that hold the fence to the table and a hole for the drill-press chuck key in the fence base, where shown.

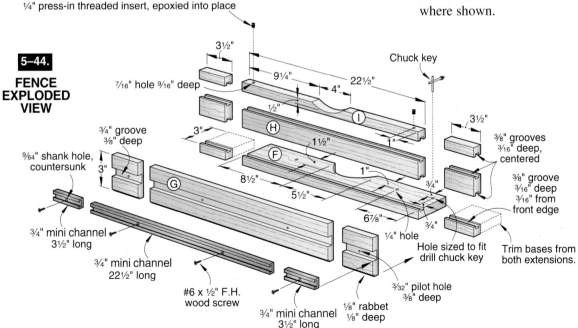

5–44.

FENCE EXPLODED VIEW

¼" press-in threaded insert, epoxied into place

3½"
9¼"
22½"
4"
½"
Chuck key

7/16" hole 9/16" deep

¾" groove ⅜" deep

3"
3"
H
I
1"
3½"

9/64" shank hole, countersunk
3"

F
1½"
8½"
1"
5½"
¾"
¾"

⅜" grooves 3/16" deep, centered

⅜" groove 3/16" deep 3/16" from front edge

G

6⅞"
¼" hole
Hole sized to fit drill chuck key

Trim bases from both extensions.

¾" mini channel 3½" long

¾" mini channel 22½" long

#6 x ½" F.H. wood screw

¾" mini channel 3½" long

⅛" rabbet ⅛" deep

3/32" pilot hole ⅜" deep

5–45. FENCE SECTION VIEW

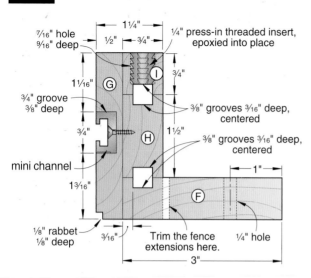

7/16" hole
9/16" deep

1 1/4"
1/2" — 3/4"

1/4" press-in threaded insert, epoxied into place

1 1/16"

3/4"

G · I

3/4" groove
3/8" deep

3/8" grooves 3/16" deep, centered

3/4"

H · 1 1/2"

3/8" grooves 3/16" deep, centered

mini channel

1"

1 3/16"

F

1/8" rabbet
1/8" deep

3/16"

Trim the fence extensions here.

1/4" hole

3"

the track in the fence aligns with the channel in the extensions, orient the flange in the same direction in all three parts.

3 Cut four, 14½"-long pieces of ⅜" steel rod with a hacksaw. Using 80-grit sandpaper, rough up 3½" at one end of each rod, and epoxy these ends into the square holes in the fence extensions. To hold the rods parallel while the epoxy cures, insert their other ends into the square holes in the fence.

4 To make knobs for locking the fence extensions in place, refer to **5–41**, on *page 140*, and thread 1½"-long flathead bolts partway into a pair of four-arm knobs. Apply epoxy under the heads, and then seat the bolts in the knobs.

5 Referring to **5–41**, slide the heads of two hexhead bolts into the auxiliary table bottom mini channel. Position the auxiliary table on the drill-press table, dropping the bolts into the slots or holes. Add washers and thread on the tapered handle knobs.

Note: The tapered handle knobs accept about ¾" of bolt length. You may need to trim the 2"-long hexhead bolts to accommodate the thickness of your drill-press table.

6 Slide hexhead bolts into the auxiliary tabletop mini channels. Align the holes in the fence base with the bolts, drop on flat washers, and fasten the fence with four-arm knobs. Slide the extension rods into the fence, and thread in the locking knobs.

6 To install press-in threaded inserts in the fence portion of part I, drill 7/16" holes to intersect the top square hole in the fence, where shown on **5–44** and **5–45**. Spread epoxy in the holes, and press the inserts in place. When the epoxy cures, ream out any excess that may have dripped into the extension rod holes with a ⅜" drill bit. For more information on using threaded inserts, see "Tips on Using Threaded Inserts" on *page 141.*

Finish and Assemble the Table and Fence

1 Cover the bottoms of the grooves and dadoes for the mini-channel in the table and fence with masking tape. Now apply a clear finish to all the parts. (We used two coats of satin polyurethane, sanding between coats with 220-grit sandpaper.) When the finish dries, remove the masking tape.

2 Using the countersunk holes in the mini channel as guides, drill pilot holes into the mating table and fence parts. Apply epoxy to the bottoms of the grooves and dadoes, and screw and clamp the mini channel in place.

*Note: The mini channel has a small flange along one outside edge, where shown on **5–45**. To make certain*

5–46.

5–47.
STOPBLOCK

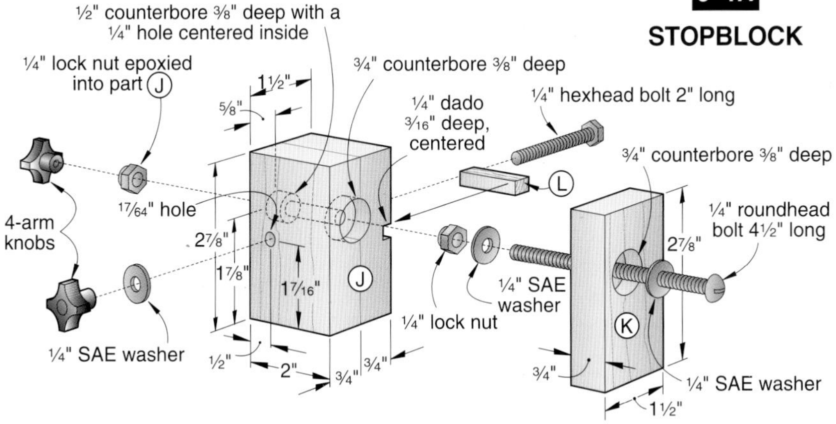
½" counterbore ⅜" deep with a ¼" hole centered inside
¼" lock nut epoxied into part J
1½"
⅝"
¾" counterbore ⅜" deep
¼" dado 3⁄16" deep, centered
¼" hexhead bolt 2" long
L
¾" counterbore ⅜" deep
17⁄64" hole
4-arm knobs
2⅞"
1⅞"
1 7⁄16"
J
¼" roundhead bolt 4½" long
2⅞"
¼" SAE washer
¼" lock nut
K
¼" SAE washer
½"
2"
¾"
¾"
¾"
¼" SAE washer
1½"

Add an Adjustable Stopblock (5–46)

1 To form the body (J), cut two pieces of ¾" stock to 2 x 2⅞", and glue them together face-to-face, keeping their ends and edges flush. With the glue dry, cut a ¼" dado 3⁄16" deep centered in the back of the body, where shown on **5–47**.

2 Cut the pad (K) to size, and adhere it with double-faced tape to the right side of the body (J) in the configuration shown on **5–47**. Chuck a ½" Forstner bit in your drill press, and drill a ⅜"-deep counterbore in the left side of the body, where dimensioned in **5–47** and as shown in **5–48**. Now, without moving the parts, change to a ¼" bit, and drill a hole, centered in the counterbore, all the way through both parts.

3 Separate the pad (K) from the body (J). Using a ¾" Forstner bit, drill ⅜"-deep counterbores in the body and pad, centered on the ¼" holes, where shown on **5–47**. To center the Forstner bit, insert pieces of ¼" dowel in the holes before drilling. Now, centering a 17⁄64" bit in the ¼" dado in the back of the body, drill a hole through the body, where shown.

4 Epoxy a lock nut in the ½" counterbore in the body (J). Then cut the guide bar (L) to size, and glue and clamp it in the dado in the back of the body, flush with its right edge, where shown on **5–47**.

5 Apply a clear finish to the parts. With the finish dry, slip a ¼" SAE washer onto the roundhead bolt, and insert it in the hole in the pad (K). (A ¼" SAE washer has an outside diameter of ⅝".) Slip another washer onto the bolt, and then thread on a lock nut. Tighten the lock nut so it firmly holds the pad, but still allows the bolt to turn. Now assemble the pad and the body (J), as shown in **5–49**, driving the bolt until the pad contacts the body.

6 Epoxy a four-arm knob onto the end of the roundhead bolt. Slide a hexhead bolt through the stop body (J) from the back, and add a washer and four-arm knob at the front, as shown on **5–47**. To use the stopblock, first adjust it to leave ½" between the pad (K) and the body. Slide the guide bar and the bolt hexhead into the mini channel, using a ruler or tape

DRILL AND ASSEMBLE THE STOPBLOCK

J
K

5–48.
With the pad (K) down and the dadoed back of the body (J) against the fence, clamp the parts in place and drill a ½" counterbore ⅜" deep in the side of the body.

J
K

5–49.
With the pad (K) mounted on the bolt with washers and a lock nut, slide the bolt into the stop body (J), and drive it into the lock nut epoxied in the body.

Materials List for Feature-Packed Table

PART	FINISHED SIZE			MTL..	QTY.
	T	W	L		
TABLE					
A base	1"	14½"	29½"	LP	1
B top sides	¼"	10⅜"	14½"	H	2
C top front	¼"	3¼"	8¾"	H	1
D top back	¼"	5¼"	8¾"	H	1
E insert	¼"	6"	8"	H	1
FENCE					
F base blank	¾"	3"	32"	M	1
G face blank	½"	3"	32"	M	1
H lower rear blank	¾"	1½"	32"	M	1
I upper rear blank	¾"	¾"	32"	M	1
STOPBLOCK					
J body	1½"	2"	2⅞"	LM	1
K pad	¾"	1½"	2⅞"	M	1
L guide bar	¼"	⅜"	1⅛"	M	1

Materials Key: LP = laminated plywood; H = tempered hardboard; M = maple; LM = laminated maple
Supplies: Masking tape; double-faced tape; ¼" dowel, epoxy.
Blades and Bits: Stack dado set; ½" and ¾" Forstner bits.
Hardware: #6 x ½" flathead wood screws (12); ¾" mini-channel: 29½" long (1), 14½" long (2), 22½" long (1), and 3½" long (2); ¼" press-in threaded inserts (2); ⅜" steel rod, 14½" long (4); ¼" flathead bolts, 1½" long (2); ¼" roundhead bolt, 4½" long (1); ¼" hexhead bolts:1½" long (2), 2" long (3); ¼" SAE flat washers (5); ¼" lock nuts (2); four-arm knobs (6); tapered handle knobs (2); ¼" ID x 1¼" OD fender washers (2); hold-down assemblies (2).

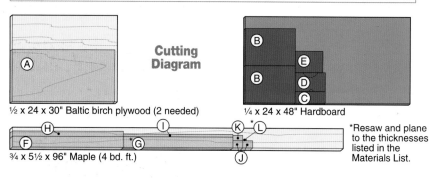

Cutting Diagram

½ x 24 x 30" Baltic birch plywood (2 needed)

¼ x 24 x 48" Hardboard

¾ x 5½ x 96" Maple (4 bd. ft.)

*Resaw and plane to the thicknesses listed in the Materials List.

measure to position the stopblock close to the desired distance from the drill bit. Clamp it in place by tightening the front knob. Now fine-tune the distance to the bit by turning the end knob. Because the clamping knob and guide bar (L) are centered in the stopblock body, you can use it on either side of the drill-press chuck by simply turning it over.

7 Assemble the hold-down clamps in the configuration shown on **5–41**, on *page 140*. Slide the hexheads of their bolts into the mini channel. Now your woodworking drill press is ready for action.

There's no need for a store-bought jig with this shop aid. We built ours (**5–50**) out of ¾" plywood and a few pieces of solid stock.

Build and assemble the jig as shown in **5–51**. Do the following:

1 Mark the hole-location centerline on the face of the workpiece, and place the workpiece against the angled fence.

2 Slide the guide assembly firmly against the workpiece, as shown in **5–52**, aligning the centerline on the guide with the one marked on the workpiece. The slots in the guide block allow you to adjust the guide for different thicknesses of wood.

3 Slide the stopblock up to the edge of the workpiece, and tighten it in place. (Because most stiles and rails require a pair of pocket holes, drill the first set of holes at one setting, loosen the wing nut, move the stopblock, and drill the second set of holes at the next setting.)

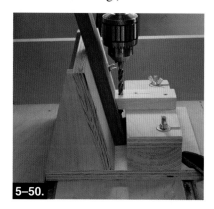

5–50.

4 Chuck a ⅜" brad-point bit into your drill press (longer bits allow more clearance). Then align the ⅜" hole in the guide with the ⅜" bit.

5 Clamp the base of the jig firmly to your drill-press table. Drill the hole into, but not through the stock, as shown on **5–52**.

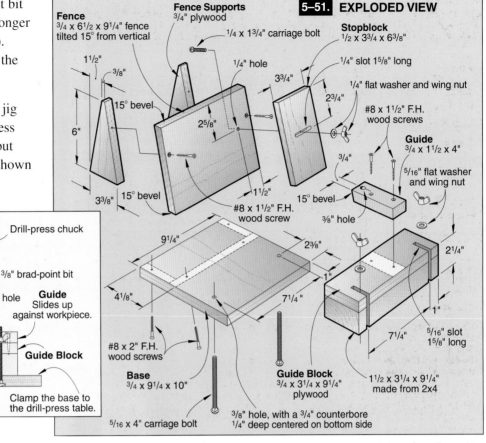

Fence
¾ x 6½ x 9¼" fence tilted 15° from vertical

1½"

3/8"

15° bevel

6"

3⅜" 15° bevel

Fence Supports
¾" plywood

¼ x 1¾" carriage bolt

¼" hole

15° bevel

2⅝"

1½"

#8 x 1½" F.H. wood screw

9¼"

4⅛"

#8 x 2" F.H. wood screws

Base
¾ x 9¼ x 10"

5/16 x 4" carriage bolt

5–51. EXPLODED VIEW

Stopblock
½ x 3¾ x 6⅜"

¼" slot 1⅝" long

3¾"

2¾"

¼" flat washer and wing nut

#8 x 1½" F.H. wood screws

Guide
¾ x 1½ x 4"

¾"

5/16" flat washer and wing nut

⅜" hole

2⅜"

7¼"

7¼"

1"

2¼"

1"

5/16" slot 1⅝" long

1½ x 3¼ x 9¼" made from 2x4

Guide Block
¾ x 3¼ x 9¼" plywood

⅜" hole, with a ¾" counterbore ¼" deep centered on bottom side

-52.

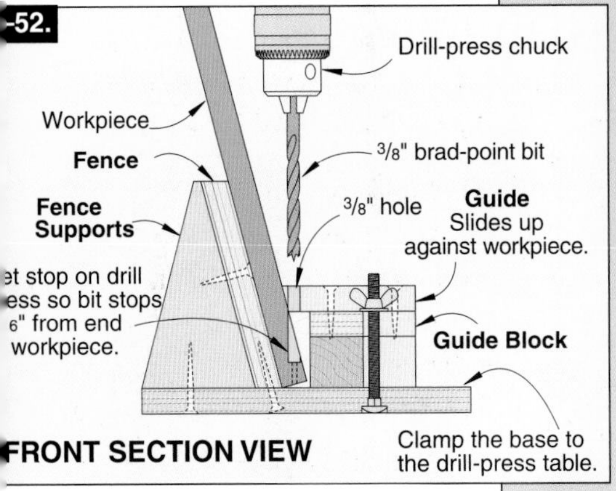

Drill-press chuck

Workpiece

Fence

⅜" brad-point bit

⅜" hole

Guide
Slides up against workpiece.

Fence Supports

et stop on drill
ess so bit stops
6" from end
workpiece.

Guide Block

Clamp the base to the drill-press table.

FRONT SECTION VIEW

6 Once the right depth has been determined, set the stop on your drill press to drill to the exact depth each time. Later, use a portable drill and a ⅛" bit to drill a pilot hole through the center of the angled ⅜" counterbore to finish creating the pocket hole for the screw.

Note: We used our jig on a heavy-duty benchtop drill press with a spindle travel of 3¼". But, we found it wouldn't work on a small benchtop model with a 2" spindle travel or models with limited clearance on the side where the drill-press stop juts out.

POCKET-HOLE JIG #2

Equipped with this handy shop aid, you can drill quick, accurate pocket holes for fastening face frames to cabinets, aprons to tabletops, and other similar tasks requiring angled mounting holes. To build the drill guide shown in **5–53**, just follow the measurements on **5–54**.

Cut a Block and Apply the Patterns

1 From 1 1/16" stock, cut a block to 2⅜ x 3½". Adhere the full-size patterns in **5–54** to it. Locate and bore a 9/16" hole,

using a spade bit in the drill press. Bandsaw the guide to shape.

5–53.

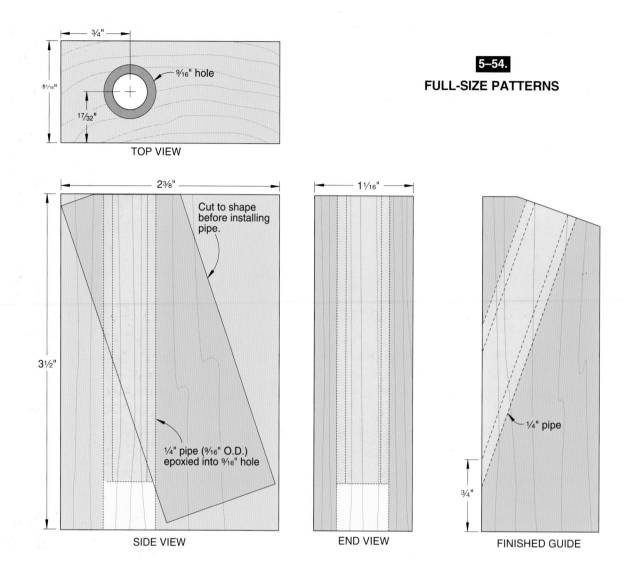

3/4"

9/16" hole

1 1/16"

17/32"

TOP VIEW

5-54.
FULL-SIZE PATTERNS

2 3/8"

Cut to shape
before installing
pipe.

3 1/2"

1/4" pipe (9/16" O.D.)
epoxied into 9/16" hole

SIDE VIEW

1 1/16"

END VIEW

1/4" pipe

3/4"

FINISHED GUIDE

2 Clamp a 3" length of 1/4" iron pipe (9/16" O.D.) into a machinist's vise. Using a 3/8" twist drill, slowly ream out the inside of the pipe to 3/8". Epoxy the pipe into the 9/16" hole, flush with the top of the block. After the epoxy sets up, hacksaw the pipe at an angle to match the block. Use a stationary sander to sand the pipe flush. Break sharp edges of steel with a file and emery cloth.

HOLD-DOWNS

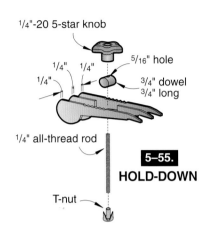

1/4"-20 5-star knob

5/16" hole

1/4" 1/4"

1/4"

3/4" dowel
3/4" long

1/4" all-thread rod

5–55.
HOLD-DOWN

T-nut

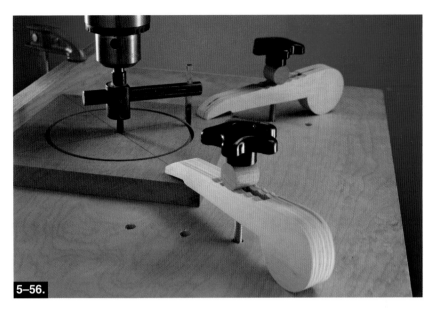

5–56.

For drilling jobs, it's essential that you hold the workpiece securely to the table and against a fence before engaging the bit. With smaller workpieces you may not have clamps with the necessary jaw depth, and, as shown in **5–56**, you don't want to get your fingers close to knuckle-busting circle cutters. Hold-downs are the answer, and **5–55** shows a version that will only set you back the cost of the knobs and all-thread rod. (Many woodworking catalogs carry such knobs in a variety of sizes.)

We drilled three holes into each side of our drill-press table for accommodating workpieces of various sizes. Each hole is outfitted with a T-nut for accepting the 1/4" all-thread rod.

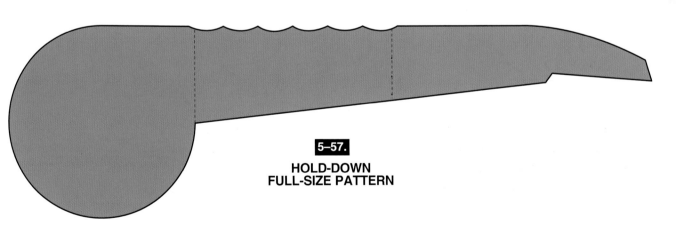

5–57.
HOLD-DOWN
FULL-SIZE PATTERN

CHAPTER 6

Sanding Solutions

LET'S FACE IT: SANDING IS A *tedious task. There's probably not a single woodworker who really enjoys it. But sanding is absolutely essential to building any project made of wood that's destined to be painted or stained and clear-finished. To help ease this necessary step, manufacturers through the years have come up with dozens of power tools, from portable belt sanders to strip sanders, profile sanders, and disc sanders, to name but a few.*

Still, there are those times when no matter how many sanding tools you own, there just isn't a single one of them that can readily perform the task at hand. So what do you do? Dog your way through it or build a jig to get it done. In this chapter, you'll find seven handy helpers to speed up some special sanding chores or, in one way or another, make it less of a hassle. Begin with the disc sander circle jig that follows, and you'll be on your way to removing the drudgery of sanding wood.

DISC-SANDER CIRCLE JIG

6–1.

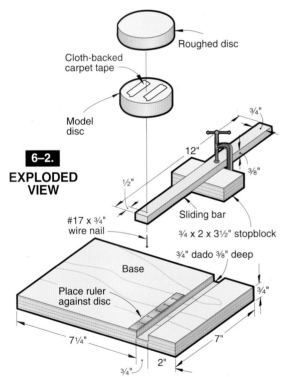

6–2.
EXPLODED VIEW

Roughed disc

Cloth-backed carpet tape

Model disc

#17 x ¾" wire nail

Sliding bar

¾ x 2 x 3½" stopblock

¾" dado ⅜" deep

Base

Place ruler against disc

12"

½"

¾"

⅜"

7¼"

7"

2"

¾"

¾"

To make the circle jig shown in **6–1** and **6–2**, do the following:

Rough-cut two circles ¼" larger in diameter than the disc you want. Then drill a hole through the center of one disc, which will serve as a model for the hole-free disc.

 Clamp a block to the sliding bar so it will stop the bar's nail at a distance from the sanding surface equal to the desired radius of your circle. Slip the model disc's hole over the nail in the bar, and sand it to size by rotating and feeding it into the sanding disc until the stopblock contacts the jig's base.

Attach the second roughed disc to the top of the model with cloth-backed carpet tape, and sand the second disc the same way you sanded the model.

RADIUS-SANDING JIG

Sanding an even radius on workpieces can be tricky, especially if you're doing it freehand. But you can take the guesswork out of this process in a hurry with this quick-fix jig (**6–3** and **6–4**). To use the jig, you'll need an oscillating spindle sander or a drum sander attached to a drill press. Do the following:

On the edge of a piece of ¾" plywood, cut out a half circle that will accommodate your largest sanding drum, as shown in **6–4**. From the edge of this half circle, measure to a point ⅛" short of the radius to be sanded, and bore a ¾" hole, where shown.

6–3.

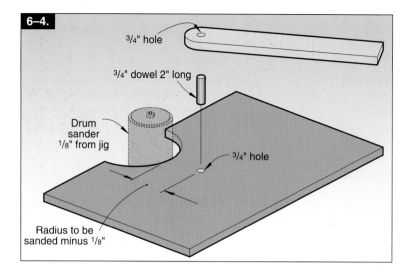

6–4.

¾" hole

¾" dowel 2" long

Drum sander ⅛" from jig

¾" hole

Radius to be sanded minus ⅛"

2 Now, glue a ¾" dowel in the hole. The accuracy of the jig depends on the dowel standing 90° to the plywood, so leave the dowel long enough to check it with a square. After the glue dries, you can cut the dowel to a shorter length.

3 Next, mark the radius on the workpiece, and cut the curve just outside the line. Bore a ¾" hole at the center point of the radius and slip the workpiece over the dowel. Adjust the plywood so the sanding drum just touches the long edge of the workpiece. When the jig is positioned correctly, clamp it to the sanding table, turn the sander on, and rotate the workpiece into the drum to sand a perfect radius.

4 If you don't want to bore a hole completely through your workpiece, you can bore the hole halfway through the stock and cut the dowel just short of this depth. For smaller workpieces, you'll want to use a dowel with a smaller diameter.

◆

RIGHT-HAND-MAN SANDING TABLE

We designed this sanding table (**6–5** and **6–6**) to fit a 10" Jet contractor's saw. You'll probably have to alter the table's size to fit your particular saw. Dust catchers like ours aren't meant to replace dust-collection devices you're already using. This one is a site-specific accessory that helps you manage the fine dust that results from using a handheld pad sander. Note also the handy built-in tool tray for storing sanding blocks, pushsticks, and other workshop items.

Build the table, as shown in **6–6**, to fit your particular saw. We used a piece of perforated hardboard to mark the numerous hole locations. Drill the holes and then countersink them slightly. Use a piece of duct out the back of the unit to fit your dust-collection system or shop vacuum.

◆

Right: A built-in dust chute links the sanding table directly to your shop vacuum or dust collector.

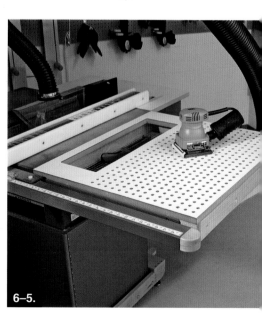

6–5.

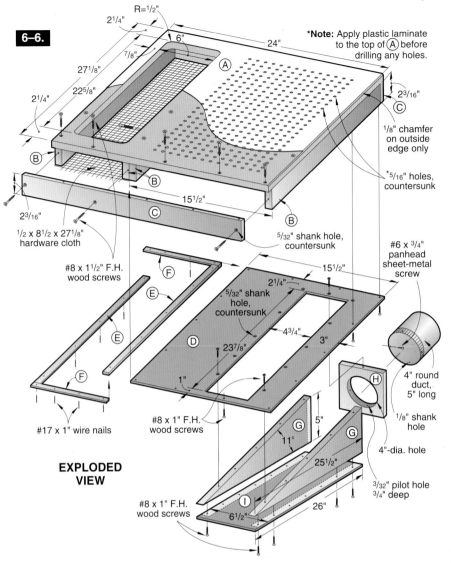

6–6.

R=1/2"
2 1/4"
6"
24"
7/8"
27 1/8"
22 5/8"
2 1/4"
2 1/4"
(A)
(C)
2 3/16"

*Note: Apply plastic laminate to the top of (A) before drilling any holes.

1/8" chamfer on outside edge only

*5/16" holes, countersunk

(B)
(B)
(B)
2 3/16"
15 1/2"
(C)
1/2 x 8 1/2 x 27 1/8" hardware cloth

5/32" shank hole, countersunk

#6 x 3/4" panhead sheet-metal screw

#8 x 1 1/2" F.H. wood screws

(F)
(E)
(E)
(F)
15 1/2"
2 1/4"
5/32" shank hole, countersunk
4 3/4"
3"
(D)
23 7/8"
1"
(H)
4" round duct, 5" long

1/8" shank hole

4"-dia. hole

#17 x 1" wire nails

EXPLODED VIEW

#8 x 1" F.H. wood screws

(G)
5"
11°
(G)
25 1/2"

#8 x 1" F.H. wood screws

(I)
6 1/2"
26"
3/32" pilot hole 3/4" deep

Sanding by hand often turns what should have been a crisp edge into one that's rounded and uneven. Reader Phil Otanicar says such round-overs are especially noticeable on the small projects he likes to make. Instead of spending money on a power edge sander, he designed a manual edge sander that clamps to his workbench (**6–7** to **6–9**).

The 12 × 18" platform supports the piece while you guide it against sandpaper attached to a fence. Three slots in the fence let you slide it up and down to expose fresh sandpaper as needed. Coarse paper is mounted on one side of the fence; the other side has finer paper. Sawdust falls into the space between the fence and platform.

Materials List for Right-Hand-Man Sanding Table

PART	FINISHED SIZE			MATL.	QTY.
	T	W	L		
SANDING TABLE					
A tabletop	3/4"	24"	27 1/8"	MF	1
B spacers	3/4"	2 3/16"	25 5/8"	H	4
C end caps	3/4"	2 3/16"	24"	H	2
D bottom	1/4"	15 1/2"	27 1/8"	HB	1
E screen molding	3/16"	3/4"	25 5/8"	H	2
F screen molding	3/16"	3/4"	8 1/2"	H	2
DUST CHUTE					
G sides	3/4"	5"	25 1/2"	H	2
H end cap	3/4"	5"	5"	H	1
I bottom	1/4"	6 1/2"	26"	HB	1

Materials Key: MF = medium density fiberboard; H = hardwood (maple or birch); H = hardboard.
Supplies: #8 x 1" flathead wood screws; #8 x 1 1/2" flathead wood screws; 1/2" hardware cloth; plastic laminate; #6 x 3/4" panhead sheet-metal screws; 4" round duct, 5" long; #17 x 1" wire nails.

6–7.

6–9. **EXPLODED VIEW**

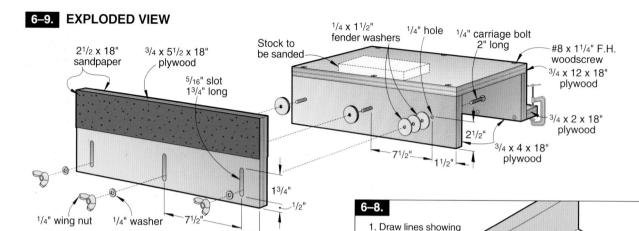

2½ x 18" sandpaper

¾ x 5½ x 18" plywood

5/16" slot 1¾" long

Stock to be sanded

¼ x 1½" fender washers

¼" hole

¼" carriage bolt 2" long

#8 x 1¼" F.H. woodscrew

¾ x 12 x 18" plywood

¾ x 2 x 18" plywood

¾ x 4 x 18" plywood

¼" wing nut

¼" washer

7½"

1¾"

½"

1½"

7½"

1½"

2½"

Make the fence first, routing the slots as shown in the four steps in **6–8**. Rout each slot in three passes—the first about ¼" deep, and each of the others about ¼" deeper than the one before. The fence should be symmetrical, so rout both end slots with the router fence and bit at the same setting. When you've finished the end slots, measure carefully, and reset the fence to rout the center slot. Once you've finished routing, lay out carriage-bolt holes in the platform using the slots as a guide. Apply a coat of gel varnish to protect the wood and reduce friction.

We bought 2½ x 180" rolls of pressure-sensitive adhesive sandpaper, cut 18" lengths, and stuck them to the fence. These rolls are available in 80 to 320 grit from woodworking suppliers.

6–8.

1. Draw lines showing edges of router bit.

5/16" straight bit

2. Draw lines showing ends of slots.

Router table

1½"

3. Align front of slot with front of router bit. Lower sander fence so it is flat on the table.

Sander fence

4. Feed forward until back of slot aligns with the back of bit. Lift fence until it is clear of bit.

PATTERN SANDER

When it came to shaping the turned-out bottoms of the legs for a nightstand being built in the *WOOD®* magazine shop, our project builder and designer gathered around the workbench for some communal headscratching. For uniformity, we needed to guide a shaping tool along a template. But a flush trimming bit chucked in our table-mounted router caused chipping where the leg curves and the wood grain intersected the routed

6–10.

6–11. EXPLODED VIEW

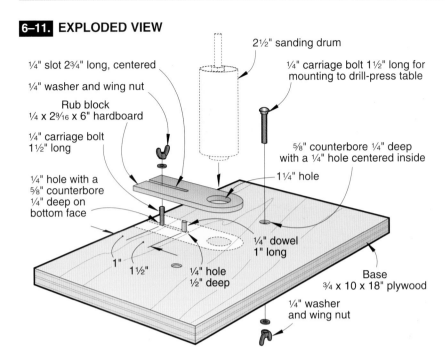

¼" slot 2¾" long, centered

¼" washer and wing nut

Rub block
¼ x 2⁹⁄₁₆ x 6" hardboard

¼" carriage bolt
1½" long

¼" hole with a
⅝" counterbore
¼" deep on
bottom face

2½" sanding drum

¼" carriage bolt 1½" long for
mounting to drill-press table

⅝" counterbore ¼" deep
with a ¼" hole centered inside

1¼" hole

¼" dowel
1" long

1"   1½"   ¼" hole
½" deep

Base
¾ x 10 x 18" plywood

¼" washer
and wing nut

face. There was also the problem of safely holding on to the narrow workpiece. Our solution? A drill-press-mounted jig that combines a "pilot-bearing" rub block with a sanding drum (**6–10** and **6–11**). Here's how to make it:

1 Cut the base from plywood or particleboard, and place it on your drill-press table, centered under the chuck. Mark the locations of your drill-press table slots, and drill counterbored holes for the mounting bolts. Drill holes for the dowel and the rub-block locking bolt. Glue the dowel in place.

2 Now, measure the diameter of your sanding drum. (You can make a rub block for each size of sanding drum you have. Ours are about ¹⁄₁₆" larger than their nominal sizes.) Cut a piece of ¼" hardboard to this width, and bandsaw and sand the radius

on one end. Rout or saw the slot, and drill the centered hole to lear the end of the sanding drum's shaft.

To use your pattern sander, fasten the base to your drill-press table with the carriage bolts. Install the rub block, and chuck up your sanding drum. Swivel the drill-press table to align the rub block side-to-side with the sanding drum, and then lock the table in place. Slide the rub block on the dowel and locking bolt to align it front-to-back. Tighten the locking bolt. Make your pattern from ½"-thick material. (We laminated two layers of ¼" hardboard.) Bandsaw your part to rough shape, and adhere your pattern to it with double-faced tape. Move the workpiece against the spinning sanding drum until the pattern contacts the rub block.

EXTRA-LONG DRUM SANDER

When our project builder started building an arched-top clock, he was faced with a plan that looked as if it would require hours of tedious hand sanding. Dreading this, he created a drill-press attachment (**6–12** and **6–13**) using parts he scrounged in the shop. The spindle itself is a length of 2" PVC plumbing pipe, a size that worked well for the clock's curves, but you can easily adapt this idea to other diameters of pipe.

Use a circle cutter to make slightly oversized wood plugs for the top and bottom of the spindle. Add the carriage bolts and nuts, chuck them into your drill press, and power-sand the plugs to fit perfectly into the pipe. Be sure that you fully countersink the

6–12.

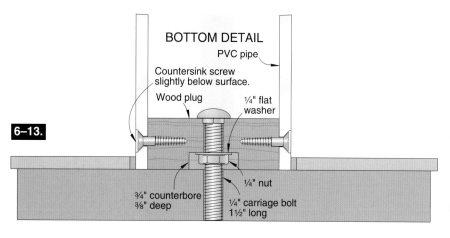

BOTTOM DETAIL

PVC pipe

Countersink screw
slightly below surface.

Wood plug

¼" flat
washer

6–13.

¼" nut

¾" counterbore
⅜" deep

¼" carriage bolt
1½" long

6–14.

PVC for the wood screws that secure the plugs so the heads don't protrude and scratch your work. Trim the sandpaper so the ends just meet or have a slight gap—overlapping the paper would make a bump.

Construct the base assembly by cutting a piece of ¼" hardboard and ¾" MDF to identical size. Drill a 2½" hole through the hardboard, and a ¼" hole through the MDF. When you set up the sander, carefully check that the spindle is square to the table. Run the sander at 500 rpm or slower. Because of the low cost of building these spindles, you may want to make one for each grit you frequently use.

❖❖❖

POSITIVE-STOP SANDING JIG

When a professional woodworker we know has a batch of parts that need to be sanded to the exact same length in a hurry, the craftsman relies on the clever sanding jig shown in **6–14** and **6–15**. "Why not just use

a power mitersaw to do the job?" you may ask. According to him, using the jig results in better grain opening, which results in better glue joints. And it's more accurate to boot, for both 90° and miter angles.

Actually, the jig couldn't be simpler. A cam-action clamp holds the workpiece tight against an auxiliary table, which slides back and forth between two

channels. Take a look at **6–15**, and you'll see that the sliding table stays snug between the channels and that precise lengths are a breeze to achieve with an adjustable stop. Pretty ingenious, don't you think?

❖❖❖

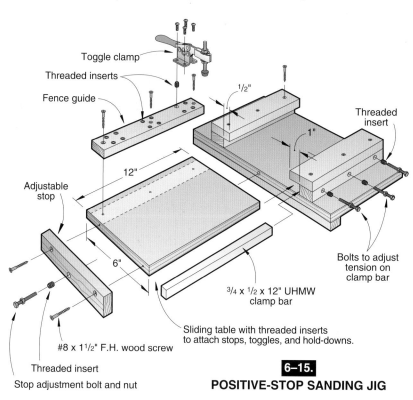

Toggle clamp

Threaded inserts

Fence guide

½"

1"

Threaded insert

12"

Adjustable stop

6"

Bolts to adjust tension on clamp bar

¾ x ½ x 12" UHMW clamp bar

Sliding table with threaded inserts to attach stops, toggles, and hold-downs.

#8 x 1½" F.H. wood screw

Threaded insert

Stop adjustment bolt and nut

6–15.
POSITIVE-STOP SANDING JIG

CHAPTER 7

Clamping Aids

NO MATTER WHAT JOINTS *you choose to use in building your woodworking projects—half-laps, dowels, dadoes, and so on—they can only do their job with the help of glues and clamps. And as you well know, there are plenty of clamp styles available in the marketplace. However, the old adage "a woodworker never has enough clamps" always seems to ring true.*

But even with all the clamps you need, there still comes those times when you need some type of clamping aid to fill the bill. You'll find some helpful ones in this chapter, as well as a super set of panel clamps.

So read on, and then get set to clamp as never before. Why not begin with the simple yet extremely useful right-angle jig on the following page?

RIGHT-ANGLE JIG

7–1.

STAY-PUT PIPE-CLAMP PADS

7–3.

Anytime you need to hold two large workpieces at a right angle, say while you're screwing or gluing them together, you need one or more of these plywood triangles. As shown in **7–1** and **7–2**, these right-angle jigs have two notches for accepting clamp jaws. The circular cutout comes in handy for temporarily holding the jig in place with a spring clamp while you position bar clamps on the notches. It also gives you a way to hang the jig on a peg when you're through with it.

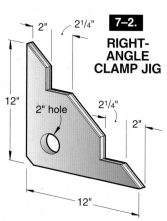

7–2.
RIGHT-ANGLE CLAMP JIG

2" 2¼"
12"
2" hole 2¼"
2"
12"

The more you use this helper, the more jobs you'll find for it. Although we designed the jig for carcase assembly, we also found it handy for holding an on-edge picture frame rigidly to a bench as we sanded the frame's edges.

Forget scrap wood. Once you try these easy-to-make pads (**7–3** to **7–5**), you'll want to crank out a pair for every pipe clamp you own.

Trying to hold a scrap wood pad in place while snugging up a pipe clamp can prove next to impossible. And, the small base on most pipe clamp also makes them prone to tipping over. Our pipe-clamp pads solve both of these annoying problems.

The pads shown here fit a Pony-brand ¾" pipe clamp, but you can adapt the design to fit other brands and sizes. For a ½" pipe clamp, bore a ⅞" hole. The size, shape, and location of the retainer notch also may vary.

To start, cut a ¾ × 3 × 24" piece of plywood, and then lay out the shape and hole locations for four pads, where dimensioned on **7–5**. Cut the pads to shape, and then drill the holes where marked. Sand or rout a slight round-over on all the edges. Cut the hardboard spacers and retainers to size, attach them with screws as shown, and you're all set to clamp down on your next project.

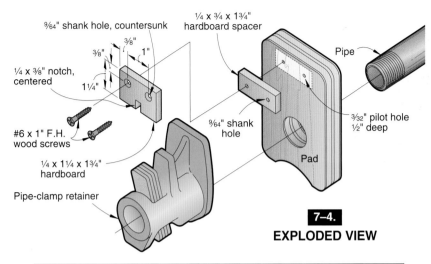

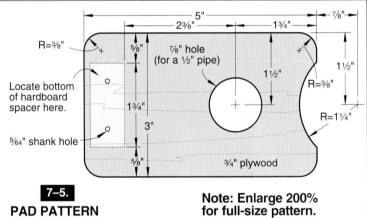

across the faces of mating boards, keeping their surfaces flush during assembly and as the glue dries. Cutouts prevent accidentally gluing the blocks to your workpieces.

To make a set of blocks, do the following:

1 Cut a 2½"-wide blank from ¾"-thick stock, as shown. Make the blank as long as you wish, working in multiple lengths of 4⅛".

2 Lay out and drill a series of 1" holes through the blank, centered on its width, where dimensioned on the drawing.

3 Rip and crosscut the blank as dimensioned. Cut a pair of ½" bevels on each block, using a bandsaw. For now, set aside the pairs you plan to use as alignment blocks. Complete the clamping blocks by adding ¼" hardboard outriggers.

PIPE-AND-BAR CLAMP BLOCKS

These easy-to-make shop aids (7–6 to 7–8) simplify edge-gluing boards. The clamp blocks spread each clamp's pressure over a wider area, and feature hardboard "outriggers" that keep the blocks in place while you position the clamps. They also prevent the clamp's pipe or bar from touching the panel's surface and creating a glue stain on your project.

The blocks without outriggers, called alignment blocks, bridge

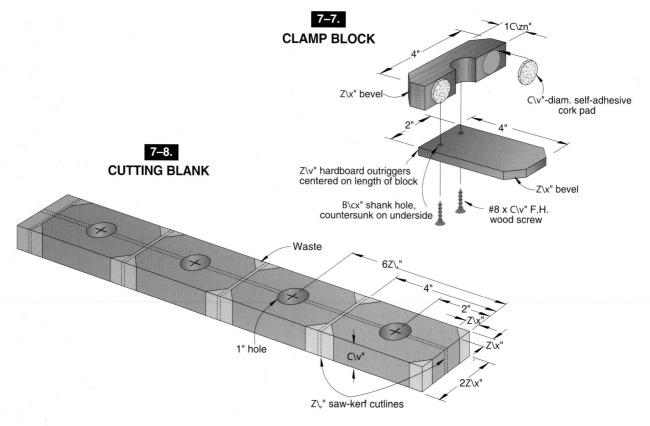

7–7.
CLAMP BLOCK

1C\zn"

4"

Z\x" bevel

C\v"-diam. self-adhesive
cork pad

2"

4"

Z\v" hardboard outriggers
centered on length of block

Z\x" bevel

B\cx" shank hole,
countersunk on underside

#8 x C\v" F.H.
wood screw

7–8.
CUTTING BLANK

Waste

6Z\v"

4"

2"
Z\x"

Z\x"

Z\x"

1" hole

C\v"

2Z\x"

Z\v" saw-kerf cutlines

4 Add a couple of coats of clear finish to all the
blocks to prevent glue from sticking to them.

5 Apply ¾" self-adhesive cork pads (available in
hardware stores) to the edges of the clamping
blocks to prevent workpiece marring.

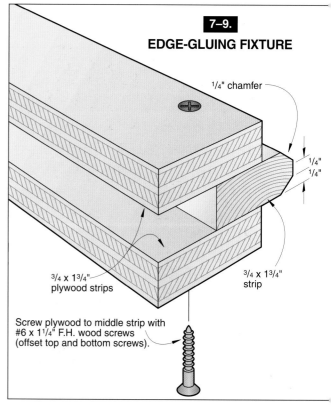

7–9.
EDGE-GLUING FIXTURE

¼" chamfer

¼"
¼"

¾ x 1¾"
plywood strips

¾ x 1¾"
strip

Screw plywood to middle strip with
#6 x 1¼" F.H. wood screws
(offset top and bottom screws).

EDGE-GLUING FIXTURE

To help clamping up a panel go smoothly, here's
an edge-gluing fixture (**7–9**) we devised in
the *WOOD*® magazine shop. It directs clamping
pressure to the center of the workpieces. Otherwise,
if the clamping pressure is not centered, it can result
in "overpull" or "underpull," which cups the panel.
You'll need two of these handy fixtures, one for each
clamp-bearing edge. As an additional advantage, this

7–10.

*Study the grain pattern and color of the boards, and then choose the most appealing arrangement. Mark a bold **V** for easy realignment later.*

fixture elevates the clamps off the panel surface, making it easier to clean up glue squeeze-out.

We also made some 3" lengths of the fixture to align the individual boards. As shown in **7–10**, you'll need two of these smaller units for each joint.

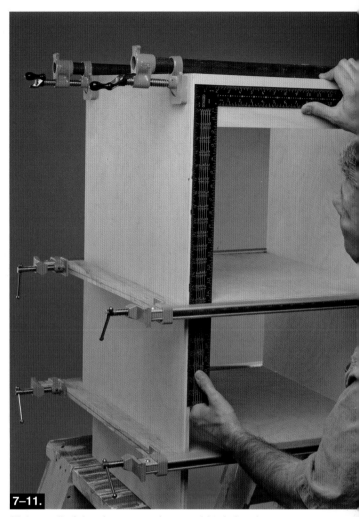

7–11.

CLAMP A LARGE BOX WITH CLAMPING CAULS

Tighten all of the clamps evenly to avoid twisting the assembly. Then check for square with a framing square and by measuring diagonally in both directions.

When building cabinets, you always want to dry-clamp the assembly first to check the fit of all the pieces, and to prepare your clamps. It all goes easier (and more accurately) if you make a pair of clamping cauls as shown in **7–11** and **7–12** for each fixed shelf in the cabinet. These cauls will help distribute clamping pressure evenly over the length of the dadoes holding these shelves. Make several; they're easy.

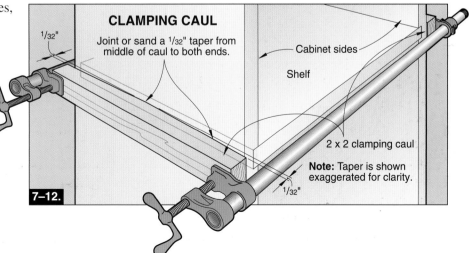

CLAMPING CAUL

1/32"

Joint or sand a 1/32" taper from middle of caul to both ends.

Cabinet sides

Shelf

2 x 2 clamping caul

Note: Taper is shown exaggerated for clarity.

1/32"

7–12.

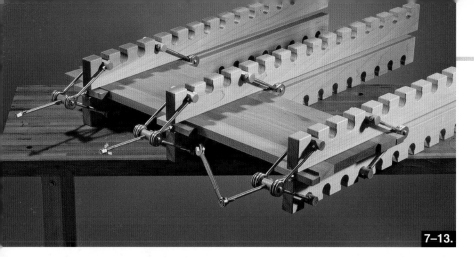

7–13.

PRESSURE-PACKED PANEL CLAMPS

These shop-built panel clamps (7–13) end edge-joining hassles. We suggest building three for starters. Do the following:

1 From 1" dowel stock, crosscut the cross dowels (A, B, C) to the lengths listed in 7–16.

2 To keep the dowels from splitting out the bottom side when drilling through them, we recommend making a drill-table support. To do this, cut two pieces of ¾ × 1 ½ × 12" stock. Rout a ½" cove along one edge of each. Glue the two pieces together in a configuration that forms a trough.

3 Center the coved assembly under a ⅝" Forstner bit chucked into your drill press. Use a stop to keep the center of the counterbores ⁹⁄₁₆" from the ends of each cross dowel (A). Drill a pair of counterbores ¼" deep into each cross dowel, as shown in 7–14 and on 7–18 on *page 162*.

4 Switch bits, and drill a ⅜" hole centered in each counterbore.

5 Repeat the process for cross dowels (B and C), centering the holes in the dowels. See 7–18 for hole sizes in the dowels.

6 Push a ⅜" T-nut into the ⁷⁄₁₆" hole in cross dowel (B) hard enough so the prongs indent the dowel. Remove the T-nut, and drill ⅛" pilot holes at each indentation. The pilot holes prevent the prongs on the T-nut from splitting the dowel later.

7 Use a rasp to form notches in the top and bottom of the cross dowel (C). Then, grind the head of the carriage bolt that will be inserted through the cross dowel (C) flush with the notches.

8 Cut the clamp pads (D) to size and shape, counterboring and drilling each, where shown on 7–18 on *page 162*.

9 Cut the clamp bars (E) to size. (We used poplar; you

7–14.

To keep the counterbores and holes aligned, we drilled the first depression and inserted a short length of ⅝" dowel stock into the counterbore. Then we aligned the dowel vertically with the bit before drilling the second counterbore.

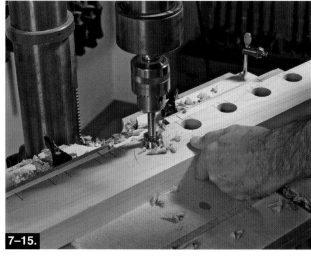

7–15.

Use a drill press with a fence and Forstner bit to drill the holes used to form the notches in the clamp bars.

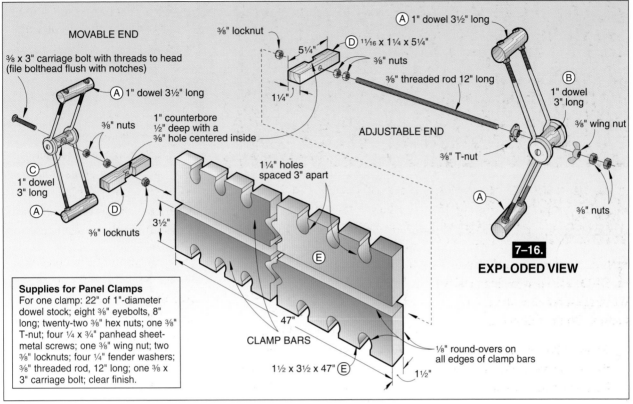

MOVABLE END

⅜ x 3" carriage bolt with threads to head
(file bolthead flush with notches)

(A) 1" dowel 3½" long

⅜" nuts

1" counterbore
½" deep with a
⅜" hole centered inside

(C)
1" dowel
3" long

(A)

(D)

⅜" locknuts

Supplies for Panel Clamps
For one clamp: 22" of 1"-diameter
dowel stock; eight ⅜" eyebolts, 8"
long; twenty-two ⅜" hex nuts; one ⅜"
T-nut; four ¼ x ¾" panhead sheet-
metal screws; one ⅜" wing nut; two
⅜" locknuts; four ¼" fender washers;
⅜" threaded rod, 12" long; one ⅜ x
3" carriage bolt; clear finish.

⅜" locknut

5¼"

(D) ¹¹⁄₁₆ x 1¼ x 5¼"

⅜" nuts

1¼"

(A) 1" dowel 3½" long

⅜" threaded rod 12" long

ADJUSTABLE END

(B)
1" dowel
3" long

⅜" wing nut

⅜" T-nut

(A)

⅜" nuts

1¼" holes
spaced 3" apart

3½"

(E)

47"

CLAMP BARS

1½ x 3½ x 47" (E)

⅛" round-overs on
all edges of clamp bars

1½"

7–16.
EXPLODED VIEW

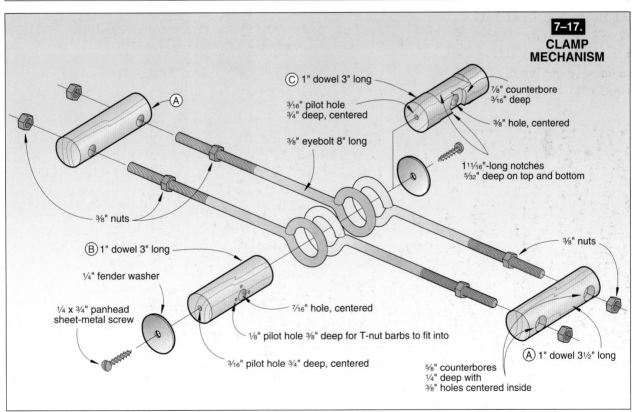

7–17.
**CLAMP
MECHANISM**

(A)

⅜" nuts

(B) 1" dowel 3" long

¼" fender washer

¼ x ¾" panhead
sheet-metal screw

(C) 1" dowel 3" long

³⁄₁₆" pilot hole
¾" deep, centered

⅜" eyebolt 8" long

⅞" counterbore
³⁄₁₆" deep

⅜" hole, centered

1¹¹⁄₁₆"-long notches
⁵⁄₃₂" deep on top and bottom

⁷⁄₁₆" hole, centered

⅛" pilot hole ⅜" deep for T-nut barbs to fit into

³⁄₁₆" pilot hole ¾" deep, centered

⅝" counterbores
¼" deep with
⅜" holes centered inside

⅜" nuts

(A) 1" dowel 3½" long

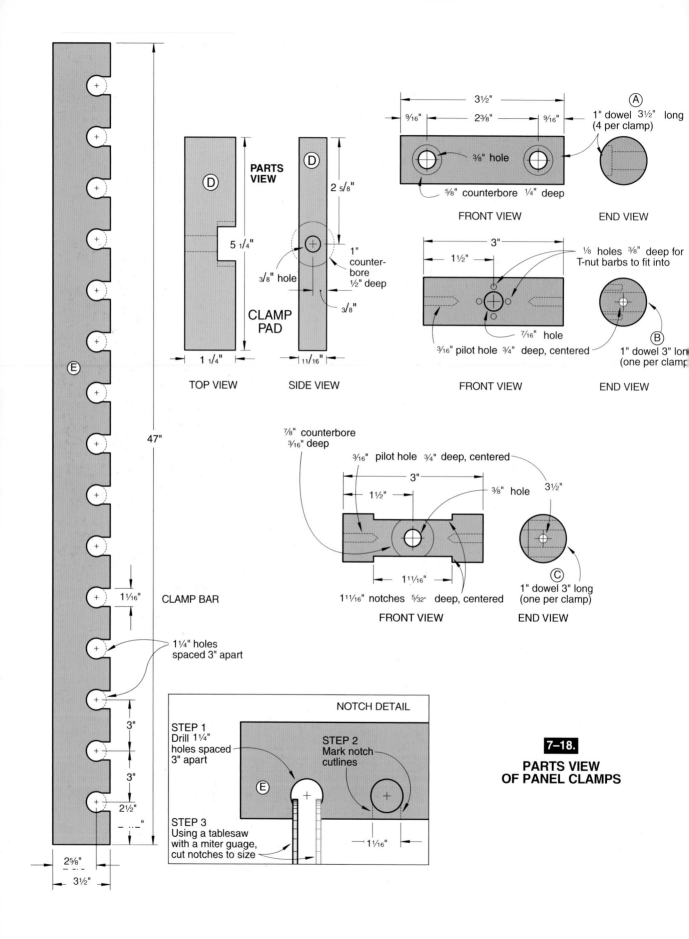

PARTS
VIEW

Ⓓ

Ⓓ

2 5/8"

5 1/4"

1"
counter-
bore
1/2" deep

3/8" hole

3/8"

CLAMP
PAD

1 1/4"

11/16"

TOP VIEW

SIDE VIEW

3 1/2"

9/16"   2 3/8"   9/16"

3/8" hole

Ⓐ

1" dowel 3 1/2" long
(4 per clamp)

5/8" counterbore   1/4" deep

FRONT VIEW

END VIEW

3"

1 1/2"

1/8 holes 3/8" deep for
T-nut barbs to fit into

7/16" hole

3/16" pilot hole 3/4" deep, centered

Ⓑ

1" dowel 3" lon
(one per clamp

FRONT VIEW

END VIEW

7/8" counterbore
3/16" deep

3/16" pilot hole 3/4" deep, centered

3"

3/8" hole

3 1/2"

1 1/2"

1 11/16"

1 11/16" notches 5/32" deep, centered

Ⓒ

1" dowel 3" long
(one per clamp)

FRONT VIEW

END VIEW

Ⓔ

47"

1 1/16"

CLAMP BAR

1 1/4" holes
spaced 3" apart

3"

3"

2 1/2"

2 5/8"

3 1/2"

NOTCH DETAIL

STEP 1
Drill 1 1/4"
holes spaced
3" apart

STEP 2
Mark notch
cutlines

Ⓔ

STEP 3
Using a tablesaw
with a miter guage,
cut notches to size

1 1/16"

7–18.
PARTS VIEW
OF PANEL CLAMPS

could also use 2 × 4 stock if you plane the edges flat and parallel.)

10 Using your drill press fitted with a fence and a Forstner bit, drill 1¼" holes 3" apart, where shown on **7–15**, on *page 160*, and **7–18**.

11 Attach an extension to your miter gauge, and make a pair of cuts at each hole to create the opening, as shown on the Notch detail in **7–18**.

12 Rout ⅛" round-overs on all edges, making the bars easier and safer to handle. Sand the bars and add a clear finish.

13 Assemble the sliding and tightening ends in the configuration shown on **7–16** and **7–17**. Keep the nuts and ends of the eyebolts flush with each other. If the bolt protrudes beyond the nut, you'll mar your workbench when using the clamps later.

14 Secure the two ⅜" nuts and wing nut to the end of the 12"-long, all-thread rod with thread lock. (We used high-strength [red] Loctite, 271.) For ease and speed in loosening and tightening the clamp ends, skip the wing nuts and use a ⁹⁄₁₆" box end wrench.

LONG-REACHING CLAMP EXTENSIONS

For clamping jobs, such as that shown in **7–19**, that require clamping pressure farther in from the edge than sliding-head clamps can accommodate, add a pair of these hardworking extensions. (We've used them on Jorgensen sliding-head-type steel bar clamps and Bessey sliding-arm bar clamps.)

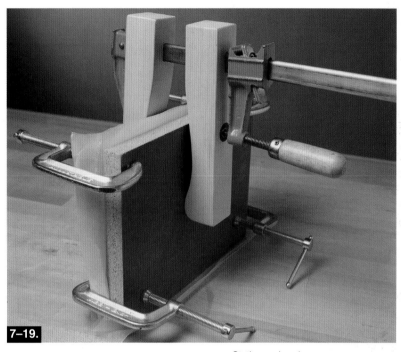

7–19.

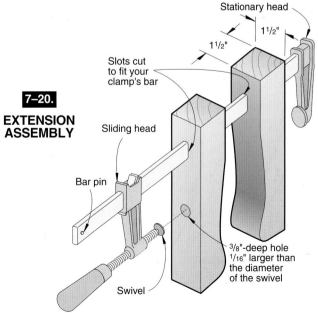

7–20.

EXTENSION ASSEMBLY

Stationary head

1½"

1½"

Slots cut to fit your clamp's bar

Sliding head

Bar pin

Swivel

⅜"-deep hole ¹⁄₁₆" larger than the diameter of the swivel

7–21.

Simply cut a pair of the extensions to shape from 1½"-square stock using **7–22** as a pattern. (We laminated two pieces of ¾" maple.) Pop the bar pin out of the end of your clamp's bar. Then, drill and cut a slot in each extension so it slides smoothly, but fits snugly on the bar.

Put the extensions on the bar in the configuration shown on **7–20**. Mark the location of the swivel on its mating extension. Remove that extension from the bar, and drill a ⅜"-deep hole ¹⁄₁₆" larger than the diameter of the swivel-head clamp end, where marked.

As shown in **7–21**, position the sliding head next to the extension when moving the two back and forth on the bar. If you leave a gap between them, they tend to bind and are harder to move in unison. Drilling the hole for the swivel in the extension allows you to slide the extension flush against the metal head.

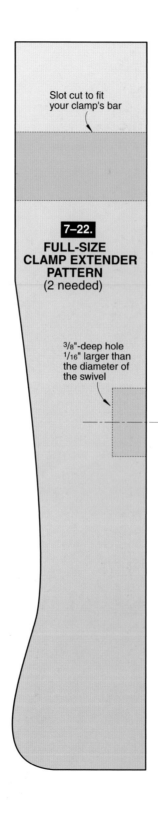

Slot cut to fit
your clamp's bar

7–22.
FULL-SIZE
CLAMP EXTENDER
PATTERN
(2 needed)

⅜"-deep hole
¹⁄₁₆" larger than
the diameter of
the swivel

ANGLE MASTER

Have you ever struggled to clamp a tapered project part, such as a triangular pediment, on top of a frame or plaque? Our wedge-shaped jig and auxiliary vise jaws (**7–23** to **7–25**) can help because the vertical dowel in the corner of the wedge pivots in the grooves. With this clamp, you can put the squeeze on a variety of angles.

To make the auxiliary jaw liners, cut two pieces of 1"-thick hardwood equal in length and width to your vise jaws. Then, put a ¼"-radius round-nose bit into your table-mounted router. Rout three equally spaced grooves across the width of the jaw liners and two grooves along the length of the jaws. Next, center the lengthwise grooves 1" from the top and bottom of the jaw liners. Now, secure the liners to the jaws.

Construct the wedge by cutting several pieces of stock using the guidelines in **7–24** for size. Glue up sufficient stock to make the depth of the wedge equal to the width of the jaw liners. Then rout the groove in the 90° corner of the wedge as shown, and fasten the dowel with glue and brads. Finally, give the wedge some gripping power by adding a piece of adhesive-backed, 100-grit sandpaper to its longest face.

7–23.

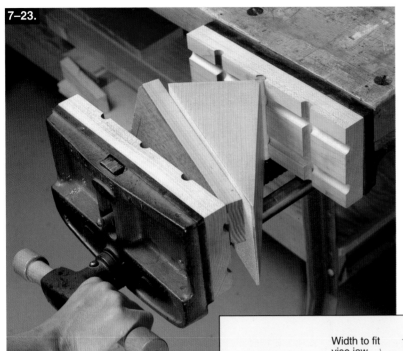

7–24. WEDGE DETAIL

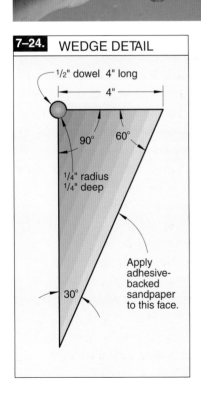

$^{1}/_{2}$" dowel 4" long

4"

90°   60°

$^{1}/_{4}$" radius
$^{1}/_{4}$" deep

30°

Apply
adhesive-
backed
sandpaper
to this face.

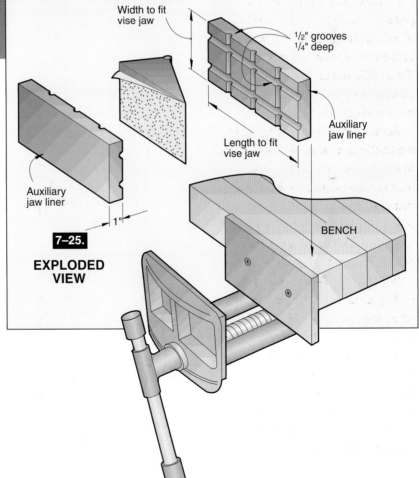

Width to fit
vise jaw

$^{1}/_{2}$" grooves
$^{1}/_{4}$" deep

Auxiliary
jaw liner

Length to fit
vise jaw

Auxiliary
jaw liner

1"

7–25.

**EXPLODED
VIEW**

BENCH

CHAPTER 8

Assorted Shop Aids

THE MOST COMMONLY AND *frequently used tools and accessories are usually situated in designated work stations or areas. Miscellaneous tools and accessories can be tucked under the workbench or stowed in a collect-all drawer. That has also been the approach taken for the chapters in this book. This chapter brings you jigs, fixtures, and accessories that just didn't seem to fit elsewhere. Take, for instance, the multipurpose thickness blocks that are discussed first. They have uses that are difficult to classify. Yet, that makes them no less handy.*

8-1.

MULTIPURPOSE THICKNESS BLOCKS

Once you make a set of hard-maple thickness blocks, you'll wonder how you ever got by without them.

Around our shop we use the blocks, shown in **8-1** and **8-2**, to set the position of fences on tablesaws, router tables, and biscuit joiners. They also come in handy for adjusting the height of sawblades and router bits.

For example, **8-1** shows how you can use them to set the depth of cut on your plunge router. Simply adjust the machine so the router bit contacts the surface the router sits on. Lock the router in place, and use the block of your choice to set the distance between the threaded

8-2.

THICKNESS BLOCKS

- 1½"
- 3"
- 1/8"
- 1/4"
- 1/2"
- 3/8"
- 5/8"
- 7/8"
- 3/4"
- 1"
- 1/4" dowel 5½" long
- 1/4" hole 3/8" deep
- 1/2"
- 2¼"
- 3¾"

depth-adjuster rod and one of the screw heads on the depth-stop turret.

We use a set of blocks that range from ⅛" to 1" thick in ⅛" increments. If you have the need, you can make thicker blocks, or produce them in finer increments, too. To make a set like ours, start with a 1 × 1½ × 36" piece of hard maple. Cut a 3"-long piece from one end for your 1" block. Then, run the workpiece through your planer until it's ⅞" thick. Saw off a 3"-long piece, and continue this process for making a block of each thickness.

With your blocks cut to thickness and length, mark each with its thickness. Drill a ⅜" hole into the blocks so they slip onto a ¼" dowel mounted to a base. You can stand the base on a work surface for portability, or attach it to a wall to save space.

❖

JIG FOR DRILLING DRAWER-PULL MOUNTING HOLES

8-3.

You can build this jig (**8-3** and **8-4**) in a minute or two, and with it you can slice many more minutes off the task of precisely drilling drawer-pull mounting holes.

To use it, first cut a ⅛" or ¼" scrap piece to 5¼ × 12". Then, mark a vertical centerline on the jig. Mark and drill centered holes for the pulls on both sides of the vertical line. For most drawers, you will want to

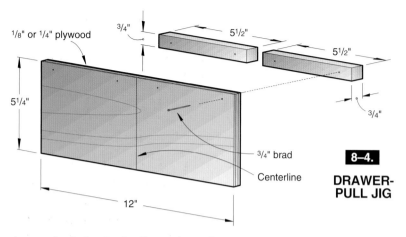

1/8" or 1/4" plywood

3/4"

5 1/2"

5 1/2"

5 1/4"

3/4"

3/4" brad

Centerline

12"

8–4.

DRAWER-PULL JIG

locate the holes in the jig so the pulls are centered on the height of the drawer fronts. Then, mark the center of each drawer front on its top edge, align the jig's vertical centerline with it, and drill the holes as shown in **8–3**.

AN EFFICIENT WAY TO INSTALL DOOR HANDLES

8–5.

8–6. THICKNESS BLOCKS

3/4"

2 3/4"

#8 x 1 1/4" F.H. wood screw

12"

12"

3 1/2"

1/8" or 1/4" plywood

This jig (**8–5** and **8–6**) works much like the drawer-pull jig *above*. But, as you can see in **8–6**, it has solid-wood cleats on both sides so you can locate handles on either the left or right side of adjoining doors. After drilling holes for one handle, just flip the jig over and drill from its other side for handles located on opposite door sides.

FAIL-SAFE HINGE JIG

Mounting hinges on an inset cabinet door is a straight-forward process. You lay the door on your workbench, locate the hinges where you want them, and screw them in place. If they need to be mortised, you mark the outline of the hinge leaf with a knife or chisel. But how do you accurately transfer the hinge locations to the carcase or to another door? *WOOD*® magazine's project builder faced this situation not once, but four times when making a set of built-in bookcases. To solve the problem, he devised a single-use jig like the one shown in **8–7** and **8–8**. To make and use the jig, do the following:

1 Cut a 2 1/4"-wide, 1/4" plywood jig rail 3" longer than the door you are hanging, and two 3/4 x 2 1/4 x 2 1/4" plywood stopblocks. Then, cut four 1/4 x 3 x 3" plywood index blocks for each hinge. Glue and clamp the stopblocks to the rail at one end, as shown on **8–8**. Now, temporarily remove the rail.

8–7.

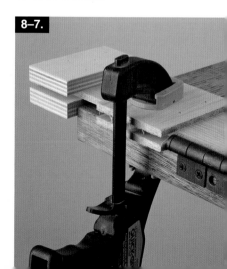

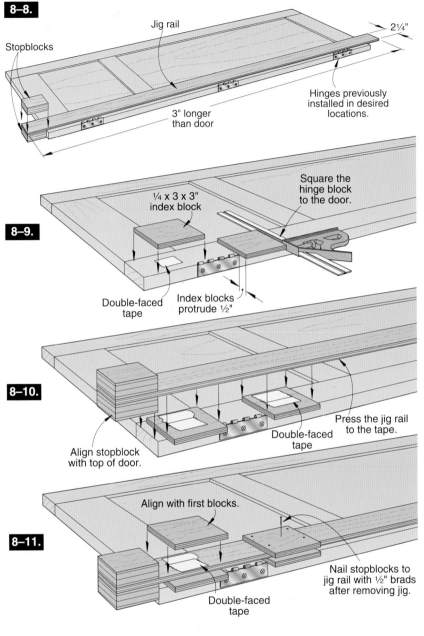

8–8.

Stopblocks

Jig rail

2¼"

Hinges previously installed in desired locations.

3" longer than door

8–9.

¼ x 3 x 3" index block

Square the hinge block to the door.

Double-faced tape

Index blocks protrude ½"

8–10.

Align stopblock with top of door.

Double-faced tape

Press the jig rail to the tape.

8–11.

Align with first blocks.

Nail stopblocks to jig rail with ½" brads after removing jig.

Double-faced tape

2 Stick small pieces of double-faced tape to the door next to the hinges. Adhere index blocks to the door, snug against the hinges, as shown on **8–9**. Stick a large piece of double-faced tape to each installed index block.

3 Now, hook one of the rail's stopblocks on the door's top edge, and press the rail onto the index blocks, as shown on **8–10**. Keep the rail's back edge and the index blocks' back edges flush.

4 To make the jig usable on both right- and left-handed doors, align a second set of index blocks with the first, sandwiching the rail between them, as shown on **8–11**. Carefully remove the rail and attached blocks from the door. Drive ½" wire brads from both sides to lock the squares in place.

5 Hooking one stopblock on each door's top edge, use the jig to locate the hinges on the rest of the doors. Drill the hinge screw pilot holes.

6 Once again, hook the jig's stopblock on a door's top edge. Mark and trim the rail to extend ¹⁄₁₆" beyond the door's

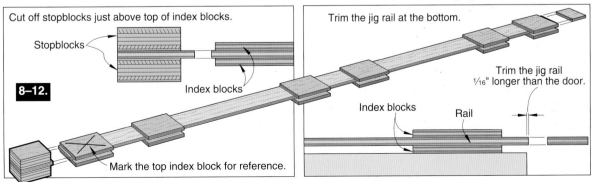

8–12.

Cut off stopblocks just above top of index blocks.

Stopblocks

Index blocks

Mark the top index block for reference.

Trim the jig rail at the bottom.

Trim the jig rail ¹⁄₁₆" longer than the door.

Index blocks

Rail

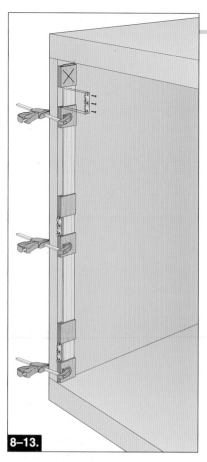

bottom, as shown on **8–12**. This extra ¹⁄₁₆" is the gap between the door and the carcase. Mark the jig's top end. Remove the stop-blocks by cutting the rail just above the top index blocks, as shown on **8–12**.

Now the jig is ready to position the hinges in the carcase. Simply place the jig against the inside of the cabinet with the rail's marked end up, as shown on **8–13**. Clamp or use double-faced tape to hold the jig in place. The index blocks bracket the hinges, just as they did on the doors. Finally, position the hinges between the index blocks, and drill the screw pilot holes.

BUILD YOUR OWN WET-WHEEL GRINDER

If you've ever turned a cutting tool blue trying to restore its edge on a bench grinder, have we got a jig for you (**8–14**)! Starting with a standard horizontal wet-wheel grinding stone and a drill press for power, we've modified the stone so it provides a steady flow of water over its surface (**8–15**). And the stone's steel backing plate rides on water-cooled plastic bearings, so the stone runs smooth and flat. We'll show you how to assemble this super sharpener for under $100, and explain how to use it, too.

Start with the Jig's Base

Cut a piece of ¾" plywood for the jig table (A) to the size listed in the Materials List on *page 172*. Mark the center-point of the table (A) and, using a

Because a stream of water constantly flows across the stone, you'll never burn a cutting edge using this jig.

compass and the centerpoint, draw a circle on the table the same diameter as the rim of the water pan. To customize the jig for your drill press, center the table (A) on your drill-press table. Mark the location of the drill-press table mounting slots, staying ¾" outside the marked diameter of the water pan.

Using your drill press, drill the ⁷⁄₆₄" pilot hole in the center of the table, ¼" mounting holes, and the holes for the tool rest support rods, where shown on the Exploded View on **8–17**.

To locate the pan's center, adjust your compass to the radius of the pan's bottom. Place masking tape across the pan's center, and with the point of the compass at the edge of the pan as shown, strike an arc. Move the compass point a third of the way around the pan and strike another arc. Then move it again and strike a third arc. The centerpoint lies at the center of the three intersecting arcs.

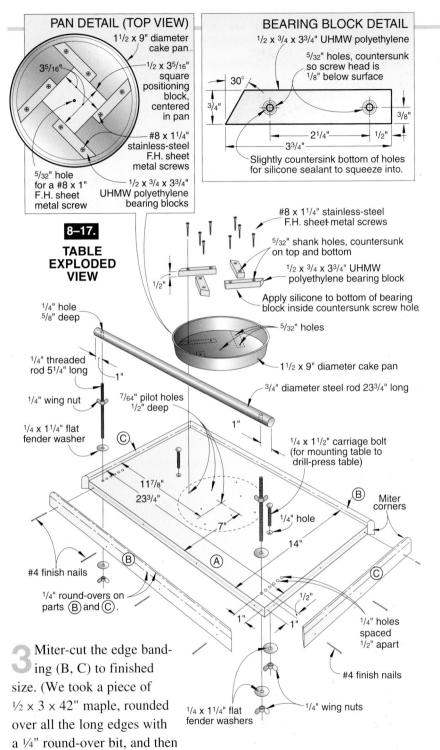

PAN DETAIL (TOP VIEW)

1¹/₂ x 9" diameter cake pan

3⁵/₁₆"

¹/₂ x 3⁵/₁₆" square positioning block, centered in pan

#8 x 1¹/₄" stainless-steel F.H. sheet metal screws

⁵/₃₂" hole for a #8 x 1" F.H. sheet metal screw

¹/₂ x ³/₄ x 3³/₄" UHMW polyethylene bearing blocks

8–17.

TABLE EXPLODED VIEW

BEARING BLOCK DETAIL

¹/₂ x ³/₄ x 3³/₄" UHMW polyethylene

⁵/₃₂" holes, countersunk so screw head is ¹/₈" below surface

30°

³/₄"

³/₈"

2¹/₄" ←→ ¹/₂"

3³/₄"

Slightly countersink bottom of holes for silicone sealant to squeeze into.

#8 x 1¹/₄" stainless-steel F.H. sheet metal screws

⁵/₃₂" shank holes, countersunk on top and bottom

¹/₂ x ³/₄ x 3³/₄" UHMW polyethylene bearing block

Apply silicone to bottom of bearing block inside countersunk screw hole.

⁵/₃₂" holes

1¹/₂ x 9" diameter cake pan

³/₄" diameter steel rod 23³/₄" long

¹/₄" hole ⁵/₈" deep

¹/₄" threaded rod 5¹/₄" long

1"

¹/₄" wing nut

⁷/₆₄" pilot holes ¹/₂" deep

¹/₂"

1"

¹/₄ x 1¹/₄" flat fender washer

Ⓒ

11⁷/₈"

23³/₄"

7"

¼ x 1¹/₂" carriage bolt (for mounting table to drill-press table)

Ⓑ

Miter corners

Ⓑ

Ⓐ

¹/₄" hole

14"

#4 finish nails

¹/₄" round-overs on parts Ⓑ and Ⓒ.

Ⓒ

1"

1"

¹/₂"

¹/₄" holes spaced ¹/₂" apart

#4 finish nails

¹/₄ x 1¹/₄" flat fender washers

¹/₄" wing nuts

3 Miter-cut the edge banding (B, C) to finished size. (We took a piece of ¹/₂ x 3 x 42" maple, rounded over all the long edges with a ¹/₄" round-over bit, and then ripped it to yield two pieces 1¹/₄" wide.) Attach the edge banding (B, C) to the table (A) with waterproof glue and nails, fill the nail holes, and finish-sand the table assembly.

4 Brush or spray a coat of oil-based primer on the entire table, and then spray on two coats of enamel. (We used an all-purpose gray primer and smoky gray spray enamel.)

Add the Water Pan Assembly

1 Cut the UHMW bearing blocks (D) to size, as shown in the Bearing Block detail in **8–17**. Drill and countersink the screw holes, where shown.

2 From ¹/₂" scrap, cut the positioning block to 3⁵/₁₆ x 3⁵/₁₆". Drill a ⁵/₃₂" hole through its center.

3 Locate the center of a 1¹/₂ x 9" aluminum cake pan, as shown in **8–16**. Drill a ⁵/₃₂" hole through the pan's center.

4 Screw the scrap wood positioning block and pan to the center of the table. Position the bearing blocks, where shown in the Pan detail in **8–17**; then drill ⁷/₆₄" pilot holes through the pan and into the table. Enlarge the pan holes to ⁵/₃₂".

5 Apply silicone sealant to the countersink on the bottom side of the bearing blocks; then fasten the blocks in place with stainless-steel screws, as shown. Remove the positioning block, apply silicone sealant to the hole, and install a panhead screw to seal the hole.

Now, Machine the Tool Rest

1 Cut a scrap piece of ³/₄" particleboard to 2¹/₂ x 23³/₄". Tilt your tablesaw blade to 45°, adjust the depth to ¹/₂", and set the fence so the tip of the

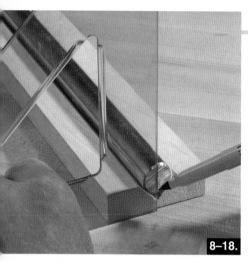

8–18.

With the ¾" steel rod mounted in the V-block, use a plastic drafting triangle or square to mark the centerline on each end of the rod.

blade rips through the center of the stock. Make one pass, rotate the stock 180°, and make a second pass to form a V-shaped notch centered on the board.

2 Cut a ¾"-diameter steel rod to 23¾" long; then smooth and

chamfer the ends with a file or grinder.

3 Apply cloth-backed, double-faced tape lengthwise near each end of the rod to keep it from rotating, and then place the rod in the V-block. Use a plastic triangle or square to mark a vertical line centered on each end of the rod, as shown in **8–18**. Extend the mark along the top of the rod.

4 With a metal punch, mark the centerpoint of the holes, where shown; then drill the holes. Remove the rod from the V-block, and set it aside.

5 Cut the ¼"-diameter all-thread support rods to 5¼" long, and file a slight chamfer on the ends. Assemble the tool rest as shown in **8–17**.

The Honing Guide Comes Next

1 Use a hacksaw, reciprocating saw, or jigsaw with a metal-cutting blade to cut the ³⁄₁₆ × 2" angle iron to 7", and then cut one side of the 2" angle to 1" wide. File or grind all the cut edges smooth, and drill and tap the holes, where shown in **8–19**.

2 Cut the ⅛ × ¾" angle iron to 7", and place it angle-side-down in the V-block you used previously. Drill a pair of ⅛" pilot holes ⅜" in from each end. Switch to a ⁹⁄₃₂" bit, and drill the holes to finished size.

3 Clean the underside of the larger angle with solvent. Apply two strips of UHMW tape to the angle, where shown in **8–19**.

8–19.

HONING GUIDE

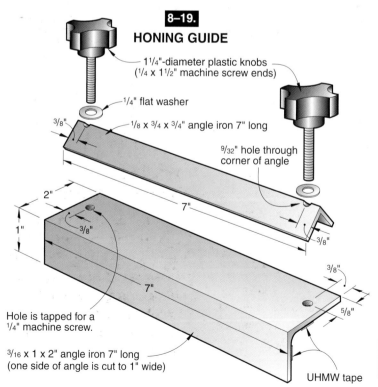

1¼"-diameter plastic knobs (¼ × 1½" machine screw ends)

¼" flat washer

⅛ × ¾ × ¾" angle iron 7" long

⁹⁄₃₂" hole through corner of angle

3/8"

2"

7"

1"

3/8"

3/8"

7"

3/8"

5/8"

Hole is tapped for a ¼" machine screw.

³⁄₁₆ × 1 × 2" angle iron 7" long (one side of angle is cut to 1" wide)

UHMW tape

Materials List for Wet-Wheel Grinder

| PART | FINISHED SIZE | | | MTL. | QTY. |
	T	W	L		
A table	¾"	14"	23¾"	P	1
B side banding	½"	1¼"	24¾"	M	2
C end banding	½"	1¼"	15"	M	2
D bearing blocks	½"	¾"	3¾"	U	4

Materials Key: P = plywood; M = Maple; U = UHMW polyethylene.
Supplies: #4 finish nails; ¾ x 23¾" steel rod; ³⁄₁₆ x 2 x 7" angle iron; ¼" carriage bolts (2), 1½" long; ¼" all-thread rods, 5¼" long (2); ¼ x 1¼" fender washers (6); ¼" wing nuts (6); #8 x 1¼" stainless-steel sheet-metal screws (8); #8 x ½" stainless-steel panhead sheet-metal screw; ½" carriage bolt, 2½" long; ⅜" flat washers (2); ½" flat washer; 1/4" flat washers (2); ½" nut; ½" coupling nut; ⅜"-drive 1¹⁄₁₆" deepwell socket; ¼ x ⅜" socket adapter; galvanized sheet metal (approx. 3 x 3"); ⁷⁄₆₄" self-tapping screw, ⅜" long; 1¼"-diameter plastic knobs with ¼" machine screw studs, 1½" long (2); ½ x 14" UHMW polyethylene self-adhesive tape; ½ x ¾ x 3¾" UHMW polyethylene blocks (4); 1½ x 9"-diameter aluminum cake pan; 200 x 25 x 75mm Makita grinding wheel (60- or 1,000-grit); silicone sealant; enamel finish; and primer.

4 Using a square and a scratch awl, scribe tool alignment marks at ½" intervals across the top face of the 1 × 2" angle, and then assemble the honing guide.

Finish Up with the Wet Wheel

(**Note:** *We looked at several different wet-wheel grinding stones before settling on Makita 60- and 1,000-grit wheels. While more expensive than most, these stones have a harder, more durable composition, and are mounted on a heavy steel backing plate.*)

1 Mark out and drill the two off-center holes in the wet-wheel backing plate, where shown on the Hole detail in **8–20**.

2 Copy **8–21**, and affix it to a piece of galvanized sheet metal. Drill the ⅛" mounting hole, and then cut the water scoop to size.

3 Cut the end of a piece of ½"-thick scrap at 30° and use it as a form to help bend the scoop to shape. Attach the scoop to the wet wheel, as shown in the Wet Wheel Exploded View in **8–20**, aligning it over the ½" hole. Assemble the ½" carriage bolt, washers, and nuts, as shown in **8–20**.

How to Use Your Wet-Wheel Grinder

Now that you've completed the jig, it's time to put a keen edge on your chisels, plane irons, and other cutting tools. But which wheel should you use, 60- or

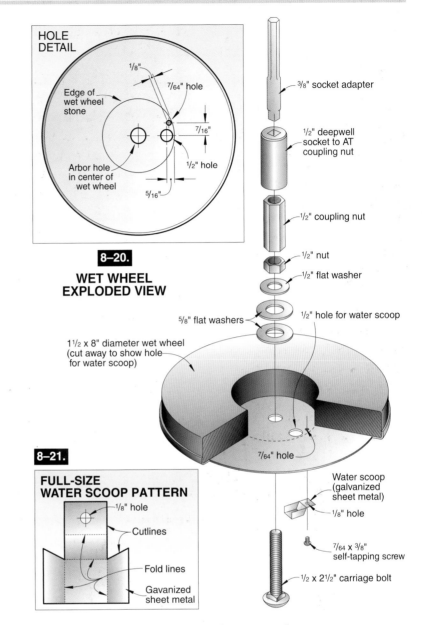

HOLE DETAIL

1/8"
7/64" hole
Edge of wet wheel stone
7/16"
1/2" hole
Arbor hole in center of wet wheel
5/16"

8–20.

WET WHEEL EXPLODED VIEW

3/8" socket adapter

1/2" deepwell socket to AT coupling nut

1/2" coupling nut

1/2" nut
1/2" flat washer

5/8" flat washers
1/2" hole for water scoop

1½ x 8" diameter wet wheel (cut away to show hole for water scoop)

7/64" hole

Water scoop (galvanized sheet metal)
1/8" hole

7/64 x 3/8" self-tapping screw

1/2 x 2½" carriage bolt

8–21.

FULL-SIZE WATER SCOOP PATTERN

1/8" hole
Cutlines
Fold lines
Gavanized sheet metal

1,000-grit? You may want to consider buying both.

We found that for general-purpose sharpening, the 60-grit wheel gave us a sharp edge quickly, making it ideal for regrinding nicked or damaged edges. For honing a fine edge on knives and carving tools, you'll need the 1000-grit wheel.

Set Up the Sharpening Jig

1 Install the jig on your drill-press table, but don't tighten the mounting screws. Center the wet wheel in the pan, and install the socket adapter and socket in the drill chuck. Raise the table and center the coupling nut under the socket, leaving a gap between the two.

8–22.

Left: *To lap a chisel, place the back side on the wet wheel with the wheel spinning away from the cutting edge as shown.*
Below: *You must lap the back side of chisels to remove factory grind marks, corrosion, and protective coatings. After lapping, the back side of the chisel should appear shiny and scratch-free.*

8–23.

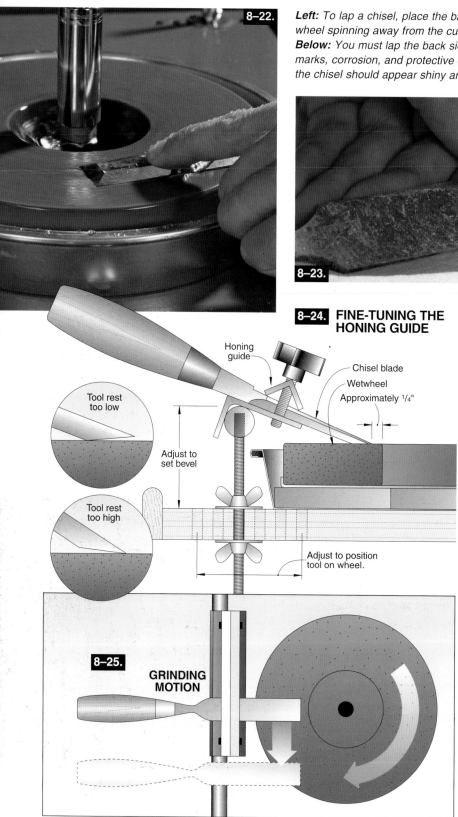

8–24. **FINE-TUNING THE HONING GUIDE**

Honing guide
Chisel blade
Wetwheel
Approximately ¹/₄"

Tool rest too low
Adjust to set bevel

Tool rest too high
Adjust to position tool on wheel.

8–25. **GRINDING MOTION**

2 Lower the quill so the coupling nut fits inside the socket, and reposition the jig so the wheel remains centered in the pan. Lock the quill and drill-press table in place, and tighten the mounting screws. To change wet wheels, simply raise the quill, swap the wheels, and then lower and lock the quill.

3 Add water to the pan until the level reaches just above the top of the bearing blocks. (Because the stone absorbs water, we like to presoak the wheel for 10 to 15 minutes.)

4 Adjust your drill-press speed to the slowest setting— usually 250 rpm—and then turn it on. The wheel should turn fast enough to pump water up from the center of the wheel and out over the stone without throwing water out of the pan. You may need to add or remove water, or adjust the drill-press speed to achieve the proper flow.

Now, Sharpen a Chisel

Before you put a fine edge on a chisel, you need to lap the back side to remove factory grind marks and flatten the surface. **Illustrations 8–22** and **8–23** show how to lap a chisel and get the results you want. With the chisel lapped, do the following:

1 Insert the chisel into the honing guide and align the edge with one of the scribed lines. Tighten the knobs to hold the chisel firmly in position.

2 Set the honing guide and chisel in place on the tool rest, as shown in **8–24**. Raise or lower the tool rest to achieve the proper edge bevel, as shown. To make sure the tool rest stays parallel to the table, measure the height at both ends, using a combination square.

3 With the tool rest adjusted and the honing guide removed, turn on the drill press. Place the honing guide on the tool rest and gently lower the chisel blade onto the rotating stone. Start with the chisel near the center of the stone, and then move it back and forth toward the edge of the wheel, as shown in **8–25**.

4 After grinding the entire bevel, carefully remove the chisel from the honing guide and wipe it dry. It will have a fine burr left from the grinding that must be removed. Simply run the cutting edge lightly across the end-grain edge of a piece of hardwood to leave a chisel that's ready for action.

Don't Forget Maintenance

Because you're introducing water into your shop, take extra time to wipe up spills and drips around your drill press. To help prevent rust, wipe down the bare metal part of your drill press with WD-40 or light machine oil each time before you use the jig.

If your grinding wheel becomes even slightly grooved, you need to flatten it. Do the following:

1 Remove the drive bolt assembly from the wheel.

2 Affix a piece of 320-grit, wet-and-dry, silicon-carbide sandpaper on a piece of glass with spray adhesive.

3 Dampen the sandpaper with water, and then rub the stone side of the wheel back and forth on the sandpaper. Rinse the sandpaper and wheel with water periodically during the process.

SIDE-TO-SIDE SHARPENING JIG

The jig shown in **8–26** and detailed in **8–27** is suitable for any chisel or plane iron with a blade at least 3" long. It's designed to sharpen tool edges at a 25° angle. You might decide to make one for each angle that you need.

To build one for yourself, do the following:

Make the Base

1 Make the base (A) from a piece of hard maple longer than the completed jig. Start with a workpiece measuring approximately ½ × 3 × 10". Install a dado set in your tablesaw, and cut a groove ³⁄₁₆" deep and 1¾" wide, ¾" from the rear edge, where shown in Step 1 of **8–28**.

8–26.

With sandpaper and this simple jig, you're only minutes away from razor-fine cutting edges.

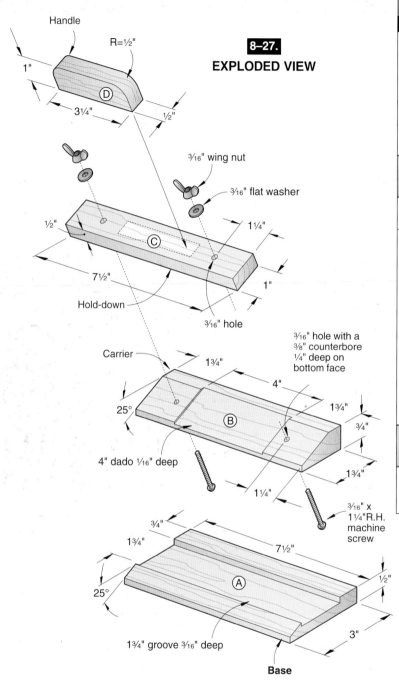

Handle

R=½"

1"

3¼"   ½"

D

8–27.
EXPLODED VIEW

³⁄₁₆" wing nut

³⁄₁₆" flat washer

½"

1¼"

C

7½"

1"

Hold-down

³⁄₁₆" hole

Carrier

³⁄₁₆" hole with a
³⁄₈" counterbore
¼" deep on
bottom face

1¾"

4"

25°

B

1¾"

¾"

4" dado ¹⁄₁₆" deep

1¼"

1¾"

³⁄₁₆" x
1¼" R.H.
machine
screw

¾"

1¾"

7½"

½"

25°

A

3"

1¾" groove ³⁄₁₆" deep

Base

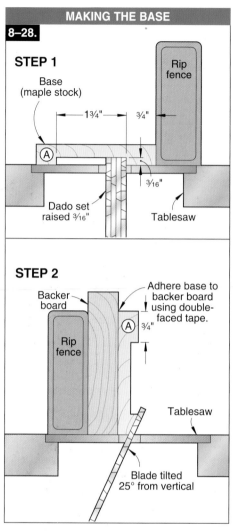

MAKING THE BASE

8–28.

STEP 1

Base
(maple stock)

Rip
fence

1¾"   ¾"

A

³⁄₁₆"

Dado set
raised ³⁄₁₆"

Tablesaw

STEP 2

Backer
board

Adhere base to
backer board
using double-
faced tape.

A   ¾"

Rip
fence

Tablesaw

Blade tilted
25° from vertical

maple. Tilt your blade to 25°, and
bevel-rip the bevel on the carrier,
again relying on a backer board
and double-faced tape, as shown
in Step 1 of **8–29**. Return the
blade to 90°, and crosscut the
carrier to its finished 7½" length.
Drill two counterbored holes from
the bottom, as shown in Step 2,
for the machine screws that
provide clamping power. Locate
each hole 1¼" from the carrier
end. Drill the counterbore first,
and then follow with the ³⁄₁₆"

2 Now, install a ripping blade,
and set it at 25°. Use double-
faced tape to fasten your workpiece to a slightly larger backer
board. Now, place this assembly
as shown in Step 2 of **8–28**, and

rip a bevel. Return the blade
to 90°, and cut the base to its
finished length of 7½".

3 Make the carrier (B) from a
piece of ¾ x 1¾ x 10" hard

MAKING THE CARRIER

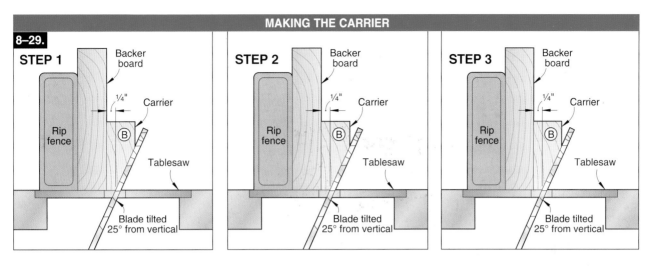

8–29.

STEP 1

Backer board
¼"
Carrier
Ⓑ
Rip fence
Tablesaw
Blade tilted 25° from vertical

STEP 2

Backer board
¼"
Carrier
Ⓑ
Rip fence
Tablesaw
Blade tilted 25° from vertical

STEP 3

Backer board
¼"
Carrier
Ⓑ
Rip fence
Tablesaw
Blade tilted 25° from vertical

8–30.

You can place the chisel against either edge of the dado. Just make sure that it's flush with that edge, flat on the carrier, and has its beveled edge flat on the work surface.

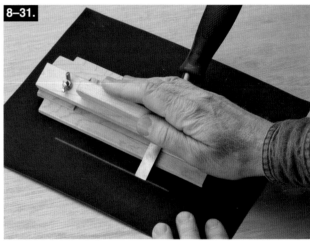

8–31.

As you begin to remove steel from the chisel, a line appears on the sandpaper. Move the jig away from the line to place the chisel on fresh abrasive. Your pressure on the jig helps keep the sandpaper in position.

through hole. Reinstall your dado set on the tablesaw and use your miter gauge, equipped with a long auxiliary fence, to cut a ¹⁄₁₆" dado 4" long, as shown in Step 3. This dado helps you clamp tools at a right angle to the work surface.

4 Cut the hold-down (C) to the dimensions shown, and drill holes for the machine screws. Locate the holes 1¼" from the hold-down ends, and centered in the width of the hold-down. Cut the handle (D), and glue it to the hold-down. After the glue dries, add the machine screws, washers, and wing nuts to make the carrier/hold-down assembly. Apply a coat of furniture paste wax to the groove in the base so that the carrier slides easily.

Refer to **8–30** and **8–31** to see how to sharpen a chisel using the jig.

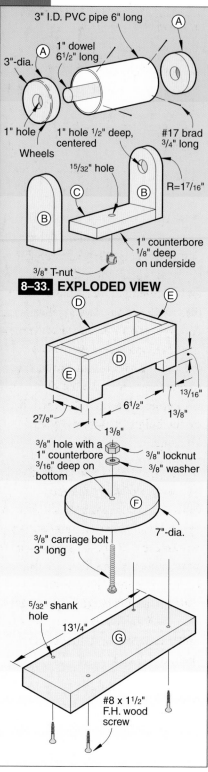

3" I.D. PVC pipe 6" long — (A)

(A) 1" dowel 6½" long

3"-dia.

1" hole

Wheels

1" hole ½" deep, centered

#17 brad ¾" long

15/32" hole

(C)

(B)

(B)

R=1⁷/₁₆

1" counterbore ⅛" deep on underside

³/₈" T-nut

8–33. EXPLODED VIEW

(D)   (E)

(D)

(E)

2⁷/₈"

1³/₈"

6½"

1³/₈"

13/16"

³/₈" hole with a 1" counterbore ³/₁₆" deep on bottom

³/₈" locknut

³/₈" washer

(F)

7"-dia.

³/₈" carriage bolt 3" long

⁵/₃₂" shank hole

13¼"

(G)

#8 x 1½" F.H. wood screw

8–32.

BENCHTOP WORK SUPPORT

If you work with benchtop tools, you know how tricky it can be trying to support long pieces of stock. This project (**8–32** and **8–33**) will end all those hassles. It supports your workpieces with precision because you can micro-adjust the height of the PVC roller through a 1" range by turning the height-adjustment disc (F).

To make the work support, do the following:

1 First, cut and epoxy the wheels (A) to the inside of the PVC roller, and insert the 1" dowel in the hole in the wheels. Sand the holes, if necessary, so the dowel rotates without binding. Glue and screw together the roller carriage (B, C) with the PVC roller in place. Then dry-clamp the base (D, E). The roller carriage (B, C) should fit snugly, but not bind inside the base. If it binds, sand the edges of the roller carriage. If it fits too loosely, make paper-thin cuts off the end grain of parts D or E, whichever needs reducing.

Materials List for Benchtop Work Support

PART	FINISHED SIZE T	W	L	QTY.
A wheels	¾"	3"-dia.		2
B roller carriage	¾"	2⅞"	5"	2
C roller carriage	¾"	2⅞"	6¼"	1
D base	¾"	3"	9¼"	2
E base	¾"	2⅞"	3"	2
F height-adjustment disc	¾"	7"-dia.		1
G sub-base	1½"	4⅞"	13¼"	1

2 Now, glue and screw the base together. Fasten the carriage bolt to the height-adjustment disc (F) with a lock nut and washer, and then thread the carriage bolt through the T-nut in the roller carriage. Insert the carriage into the base, and screw the base to the sub-base (G). You may need to make the sub-base out of thicker or thinner stock depending on the height of the tools you'll be using.

3 To get the roller dead level with your tool tabletop, position both on your bench the appropriate distance apart for the workpiece. Clamp the support to your benchtop, and lay a straight-edge across the tabletop and roller. Turn the height-adjustment disc until you can't see any light underneath the straightedge on the tool tabletop.

8–34.

AUXILIARY PLANER BED

Thickness planers are wonderful tools, but like most machines they have their limitations. Many of them will not thickness stock thinner than ¼", and none of them will plane a bevel. However, outfitted with an auxiliary bed, any planer can perform both of these tasks. We designed our bed to fit a Delta portable planer, but you can change the dimensions, and the shape of the bottom-side cleats, to fit any planer (**8–34** and **8–35**). (The cleats prevent the jig from sliding on the table.)

The ¾" thickness of the auxiliary bed raises the height of the workpiece so you can plane stock less than ¼" thick. There's no danger of damaging the knives because at the worst they will only cut slightly into the plywood surface.

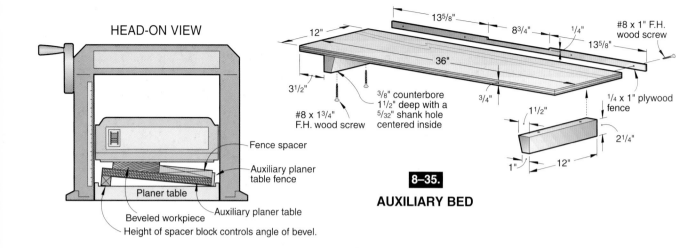

HEAD-ON VIEW

Fence spacer

Auxiliary planer table fence

Planer table

Beveled workpiece

Auxiliary planer table

Height of spacer block controls angle of bevel.

13⁵/₈"

12"

8³/₄"

1/4"

#8 x 1" F.H. wood screw

36"

13⁵/₈"

3¹/₂"

#8 x 1³/₄" F.H. wood screw

³/₈" counterbore 1¹/₂" deep with a ⁵/₃₂" shank hole centered inside

³/₄"

1¹/₂"

¹/₄ x 1" plywood fence

2¹/₄"

1"

12"

8–35.

AUXILIARY BED

By adding a spacer block under one side of the auxiliary bed, as we did in **8–34** and **8–35**, you can raise that side so the planer cuts a bevel. You'll find this handy for making such things as siding and thresholds. Note in **8–34** that we clamped down both sides of the jig. And, we added a fence spacer on the low side of the jig to prevent the planer knives from cutting into the bed before it cuts the full depth of the bevel.

Smooth hardwood plywoods, such as birch or maple, work well for the auxiliary bed. If you use a lesser grade, sand it as smooth as possible and apply paraffin wax to lubricate its surface.

8–36.

RADIAL-ARM-SAW AUXILIARY TABLE

If you're fortunate enough to have a radial-arm saw that you can dedicate to making right-on 90° cuts, here's an idea for you (**8–36**). Take as much time as you need to adjust your saw so it's perfectly perpendicular to the saw's fence. Now, lock it in place. Then, if you need to make dadoes, grooves, or rabbets, simply install your dado set, position the auxiliary table atop the saw table, raise the blade to the appropriate height, and make your cuts.

When you're finished, remove the auxiliary table, change blades, and lower the blade back down for your next cutting operation.

METRIC EQUIVALENTS CHART
Inches to Millimeters and Centimeters

MM=MILLIMETERS			CM=CENTIMETERS			

INCHES	MM	CM	INCHES	CM	INCHES	CM
1/8	3	0.3	9	22.9	30	76.2
1/4	6	0.6	10	25.4	31	78.7
3/8	10	1.0	11	27.9	32	81.3
1/2	13	1.3	12	30.5	33	83.8
5/8	16	1.6	13	33.0	34	86.4
3/4	19	1.9	14	35.6	35	88.9
7/8	22	2.2	15	38.1	36	91.4
1	25	2.5	16	40.6	37	94.0
1 1/4	32	3.2	17	43.2	38	96.5
1 1/2	38	3.8	18	45.7	39	99.1
1 3/4	44	4.4	19	48.3	40	101.6
2	51	5.1	20	50.8	41	104.1
2 1/2	64	6.4	21	53.3	42	106.7
3	76	7.6	22	55.9	43	109.2
3 1/2	89	8.9	23	58.4	44	111.8
4	102	10.2	24	61.0	45	114.3
4 1/2	114	11.4	25	63.5	46	116.8
5	127	12.7	26	66.0	47	119.4
6	152	15.2	27	68.6	48	121.9
7	178	17.8	28	71.1	49	124.5
8	203	20.3	29	73.7	50	127

BOARD FEET GUIDELINES

Dealers typically price softwoods by the running (lineal) foot and hardwoods by the board feet (a volume measurement). A board foot includes thickness, width, and length measurements that equal 144 cubic inches. See "How to Determine Board Feet" below to see some sample calculations.

The thickness of lumber, especially hardwoods, is referred to in quarters of an inch, such as 4/4 ("four/quarters" or 1"), 5/4 (1.25"), 6/4 (1.5"), 8/4 (2"), and so on. However, these hardwood thicknesses are designated and the board footage calculated before surfacing. Although you'll pay for the full designated thickness, what you'll actually get in lumber surfaced two sides (S2S) is shown in "Effects of Planing" below. Also, in a lumber store the board footage is rounded up or down to the nearest one-half board foot, except for more costly exotic or imported wood. Exotic wood is calculated to the inch, and will be rounded to the nearest hundredth of an inch.

HOW TO DETERMINE BOARD FEET
Whether you measure a board's length in inches or feet, calculating board footage for that piece of lumber is simple math:

$$\frac{\text{T x W X L}}{144} = \text{board feet}$$

$$\frac{\text{T x W X L}}{12} = \text{board feet}$$

1' x 6' x 96' = 576
576 ÷ 144 = 4 board feet

1' x 6' x 8' = 48
48 ÷ 12 = 4 board feet

EFFECTS OF PLANING

When you buy lumber, you pay for its rough thickness before surface-planing on two sides (S2S). Here's how planning affects the thickness of purchased boards. You can save money by buying full-thickness rough stock from mills that offer it, and then plane it yourself.

HARDWOOD THICKNESS AFTER S2S	LUMBER ROUGH THICKNESS
13/16"	4/4 = 1"
11/16"	5/4 = 1¼"
15/16"	6/4 = 1½"
1¾"	8/4 = 2"
2¼"	10/4 = 2½"
2¾"	12/4 = 3"

INDEX

CREDITS

Special thanks to the following people or companies for their contributions:

Marty Baldwin, for photographs in Chapters 4, 7, and 8

Baldwin Photography, for photographs in Chapters 2, 3, 5, and 6

Jim Boelling, for text in Chapter 2 and project design in Chapter 3

Kevin Boyle, for text in Chapter 3

David Brunson, for project design in Chapter 3

Thomas Bruzan, for project design in Chapter 2

Tim Cahill, for illustrations in Chapter 3

Dave Campbell, for text in Chapter 3

Larry Clayton, for text in Chapter 3

Jeff Day, for text in Chapters 2 and 6

James R. Downing, for project designs in Chapters 2, 3, 5, 6, and 8 and text in Chapters 2 and 3.

Kim Downing, for illustrations in Chapters 3, 4, 5, 7, and 8

Owen Duvall, for text in Chapters 4 and 5

Gary Elderton, for project design in Chapter 8

Charles I. Hedlund, for project designs in Chapters 2, 3, 4, and 7 and text in Chapters 2, 3, and 4

John Hetherington, for photographs in Chapters 2 to 8

William Hopkins, for photographs in Chapters 3 and 5

Tom Jackson, for text in Chapters 2, 3, and 4

Brian Jensen, for illustrations in Chapter 3

Dan Johnson, for project design in Chapter 7

Lorna Johnson, for illustrations in Chapters 2 to 8

Marlen Kemmet, for text in Chapters 2, 3, 4, 5, and 7

King Au, for photographs in Chapters 6 and 7

Bill Krier, for text in Chapters 2 and 3

Erich Lage, for illustrations in Chapter 8

Bill LaHay, for text in Chapter 5

Roxanne LeMoine, for illustrations in Chapters 2, 3, 4, 5, 6, 7, and 8

Perry McFarlin, for graphic design in Chapter 8

Jeff Mertz, for project designs in Chapters 2 and 7 and text in Chapters 3, 4, and 5

Elton Miller, for text in Chapter 3

Mike Mittermeier, for illustrations in Chapter 6

Steve Oswalt, for text in Chapter 8

Phil Otanicar, for project design in Chapter 6

Jim Pollock, for text in Chapters 3, 4, and 8

Erv Roberts, for project design in Chapter 2

Mike Sarnes, for project design in Chapter 7

John Schlabaugh, for text in Chapter 3

Studio Au, for photographs in Chapter 2

Jan Svec, for text in Chapters 2, 4, 5, 6

Dan Tanner, for photographs in Chapter 7

Robert Taugher, for project design in Chapter 5

Carl Voss, for text in Chapter 2

Raymond L. Wilber, for text in Chapter 2

Sam Williams, for text and illustrations in Chapter 2

Joe Xavier, for project design in Chapter 3

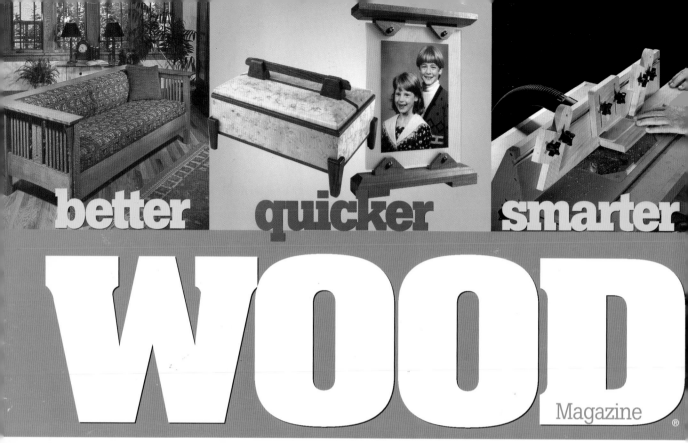